Advancing Combat Support to Sustain Agile Combat Employment Concepts

Integrating Global, Theater, and Unit Capabilities to Improve Support to a High-End Fight

JAMES A. LEFTWICH, KATHERINE C. HASTINGS, VIKRAM KILAMBI,
KRISTIN F. LYNCH, SHANNON PRIER, RONALD G. MCGARVEY

Prepared for the Department of the Air Force
Approved for public release; distribution unlimited

For more information on this publication, visit **www.rand.org/t/RRA1001-1**.

Published by the RAND Corporation, Santa Monica, Calif.

Library of Congress Cataloging-in-Publication Data is available for this publication.

ISBN: 978-1-9774-1105-1

Cover: U.S. Air Force photo by Senior Airman Jonathan Valdes Montijo.

About This Report

To improve operational resiliency when facing increasingly capable adversaries, the U.S. Air Force is evolving its fundamental concept of employment from one focused on static force beddowns at established forward bases, which is heavily reliant on robust missile defense capabilities, to a maneuver concept that seeks to reduce the effectiveness of adversary attack by complicating enemy targeting. Conducting the types of maneuvers envisaged in the evolving employment concept requires a dynamic and agile logistics and sustainment network. The objective of this research was to evaluate the combat support (CS) enterprise from the global, theater, and base levels, as well as explore various approaches for providing logistics and sustainment support for combat forces conducting maneuver operations.

This report summarizes the findings of the first part of this research project, which focused on decomposing the CS enterprise to understand how decision authorities and CS resource characteristics influence options for a theater-level logistics network design. It should be of interest to the CS community, specifically those responsible for supporting theater-level flying operations in competition and conflict. A companion report, which is not available to the general public, provides our analysis of the effectiveness of different designs in supporting combat operations. This report specifically looks at the value of different munitions loadouts and the number of additional days of operations provided by alternative loadouts, as well as the value of investments in buffer stocks or spare parts to remedy aircraft availability when spare parts availability becomes a limiting factor.[1]

The research reported here was commissioned by the Director of Logistics, Engineering, and Force Protection, Headquarters, Pacific Air Forces, and conducted within the Resource Management Program of RAND Project AIR FORCE as part of fiscal year 2021 project, "Linking Enterprise, Theater, and Wing-Level Combat Support in a High-Threat Environment."

RAND Project AIR FORCE

RAND Project AIR FORCE (PAF), a division of the RAND Corporation, is the Department of the Air Force's (DAF's) federally funded research and development center for studies and analyses, supporting both the United States Air Force and the United States Space Force. PAF provides the DAF with independent analyses of policy alternatives affecting the development, employment, combat readiness, and support of current and future air, space, and cyber forces.

---

[1] Katherine C. Hastings, James A. Leftwich, Vikram Kilambi, and Ronald G. McGarvey, *Maneuvering Beyond Support: Application of the PLATO Model to Determine Sufficiency of Munitions, Fuel, and Spare Parts Required to Support a Pacific Maneuver Operation*, RAND Corporation, forthcoming, Not available to the general public.

Research is conducted in four programs: Strategy and Doctrine; Force Modernization and Employment; Resource Management; and Workforce, Development and Health. The research reported here was prepared under contract FA7014-16-D-1000.

Additional information about PAF is available on our website:

www.rand.org/paf/

This report documents work originally shared with the DAF on September 30, 2021. The draft report, issued on September 30, 2021, was reviewed by formal peer reviewers and DAF subject-matter experts.

Acknowledgments

We thank Brigadier General Sean Tyler and Jim Silva for their initiation of this work and support throughout its execution. We also thank Colonel Daniel Lockert for his continued support during the project.

We thank the numerous individuals in the U.S. Air Force who shared information and insights with us. Although too numerous to list individually, personnel on the Headquarters, Pacific Air Forces staff in the A4, A5/8, A3/6, and A9 directorates were particularly helpful in providing data and key insights on theater operations and support plans—the very context of our analysis. Other personnel in various organizations were instrumental in providing data to support our analysis. We thank the following organizations for their help: the Air Force Sustainment Center (Global Ammunition Control Point and the 635th Supply Chain Operations Wing), the Air Force Life Cycle Management Center (F-15 System Program Office and F-16 System Program Office), the Office of the Deputy Assistant Secretary of the Air Force for Operational Energy, and the Air Force Materiel Command Analysis Division (specifically Rich Moore, who frequently engaged to share insight on data and ongoing Air Force initiatives).

At RAND, we thank Kristin Van Abel, Rachel Costello, Christopher Lynch, Andrew Karode, Aimee Bower, and Patrick Mills for their support. We also thank Brent Thomas and Jeremy Eckhause for their thoughtful reviews that improved the quality of this report, as well as Anna Jean Wirth and again Jeremy Eckhause who reviewed the model methodology used to support our analysis.

That we received help and insights from those acknowledged above should not be taken to imply that they concur with the views expressed in this report. We alone are responsible for the content, including any errors or oversights.

Summary

For many years, absent a peer adversary and in the face of tightening budgets, the joint logistics enterprise moved from a focus on effectiveness (i.e., where the priority is enabling combatant command combat operations, with less attention on costs and resource utilization) to a focus on efficiency.[2] The focus on efficiency has driven peacetime logistics and sustainment processes to be more centralized in the U.S. Air Force (USAF) and, in some cases, at the U.S. Department of Defense level. In some instances, the centralization placed decision authorities associated with the allocation and reallocation of resources outside the control of warfighting commands. Additionally, the move toward efficiency has created a lean supply chain that relies on assured transportation to rapidly deliver resources where needed based on demand signals from end-users. Capable adversaries, however, can disrupt the supply chain by degrading communications and limiting access to forward locations.

Motivation and Approach

As Pacific Air Forces (PACAF) pursues evolving operational concepts of employment (CONEMPs) designed to improve operational resiliency, questions about the fragility of the combat support (CS) enterprise persist. How should the CS enterprise operate if communication networks are degraded and requests for resources cannot be transmitted to supply sources? What actions can be taken to mitigate against uncertainties in supply chain performance in contested environments? In light of these questions, Headquarters PACAF asked RAND Project AIR FORCE researchers to assess the CS enterprise holistically, including base, theater, and global resources, and explore different concepts that could be integrated in theater sustainment plans to support operations.

Our analysis was conducted in two parts: (1) decomposing the CS enterprise from decision authority and resource characteristic perspectives and (2) analyzing differences in CS enterprise posture performance in the context of the evolving operational CONEMPs.[3] The first part of the analysis (covered in this report) informs a framework that PACAF can use to consider the

---

[2] Joseph F. Dunford, Jr., "Institute for Defense Analysis Study," information memorandum to the Secretary of Defense, April 30, 2018.

[3] The CS enterprise performance in the context of evolving operational CONEMPs is addressed in a companion report, which is not available to the general public; that report illustrates how a range of maneuver schemes that might be contemplated in competition and in conflict could be better supported by establishing buffer stocks or considering alternative munitions loads (Katherine C. Hastings, James A. Leftwich, Vikram Kilambi, and Ronald G. McGarvey, *Maneuvering Beyond Support: Application of the PLATO Model to Determine Sufficiency of Munitions, Fuel, and Spare Parts Required to Support a Pacific Maneuver Operation*, RAND Corporation, forthcoming, Not available to the general public).

necessary elements of the CS enterprise for operating in a hybrid push-pull system as a means to mitigate uncertainty and adversary actions that challenge logistics support. This report also presents the cost of various resource buffer strategies for spare parts.

Key Findings

Our analysis revealed the following:

- The CS enterprise, built on efficiency, relies on timely communication of asset requirements and on assured and responsive transportation, which today fails to achieve desired readiness rates and will likely be more challenged in a conflict with a near-peer adversary.
- Processes for communicating resource status and replenishment needs will aid in mitigating CS enterprise disruptions resulting from adversary attacks; however, other actions will likely be necessary to support operational missions in a contested environment.
- Operating the CS enterprise in conflict will require logisticians above the unit level to be fully aware of planned sortie demand and able to forecast required replenishment based on that demand.
- Planning factors used to make posture decisions during competition may not accurately reflect the expected intensity of operations in a conflict with a near-peer adversary.

Recommendations

In light of our findings from this analysis, we recommend that USAF consider the following mitigation strategies:

- Each PACAF CS functional area, in coordination with its enterprise functional community, should develop and practice tactics, techniques, and procedures for executing the logistics support plan in a communications-degraded environment.
- PACAF should adopt a methodology to determine which assets in the CS enterprise should be pushed to forward operating locations and the rate and quantities that should be pushed if it becomes necessary to temporarily shift from a pull to a push system in a communications-degraded environment. We propose the basis for such a methodology in Chapter 4 of this report.
- USAF and PACAF should consider a "partial buyout" buffer stock strategy to mitigate expected resource shortages for planned operations.
- USAF should review the War and Mobilization Plan–Volume 5 planning factors used to compute readiness spares packages to ensure these factors reflect the intensity of operations outlined in PACAF's operations plan, including expected attrition.

Contents

Figures

Figures

Chapter 1. Background and Motivation

For many years, absent a peer adversary and in the face of tightening budgets, the joint logistics enterprise (JLEnt) moved from a focus on effectiveness (i.e., where the priority is enabling combatant command combat operations with less attention on costs and resource utilization) to a focus on efficiency.[4] The focus on efficiency has driven peacetime logistics and sustainment processes to be more centralized in the U.S. Air Force (USAF) and, in some cases, at the U.S. Department of Defense (DoD) level. Management of resources varies by commodity and function; some are managed centrally, and others are decentralized. The centralized management of resources can result in uncertainty about resource availability and responsiveness of the combat support (CS) enterprise to meet dynamically changing theater wartime requirements.[5]

Additionally, the United States faces adversaries that are increasingly more capable of putting combat and support forces at risk of kinetic and non-kinetic attack and who can restrict or deny access to the battlespace. In response to improved adversary capabilities, the USAF operational community has evolved concepts of employment (CONEMPs) designed to increase operational resiliency. These new CONEMPs place additional demands on resources, the responsiveness of the CS enterprise, and the ability of the CS enterprise to operate proactively and perhaps independently of traditional lines of communication.

---

[4] As defined in Joint Publication 4-0, the JLEnt "is a multitiered matrix of key global logistics providers cooperatively structured through an assortment of collaborative agreements, contracts, policy, legislation, or treaties utilized to provide the best possible support to the JFC [joint force commander] or other supported organization. The key DOD organizations in the JLEnt include the Services, combatant commands (CCMDs), Defense Logistics Agency (DLA), Joint Staff J-3 [Operations Directorate], and Joint Staff J-4 [Logistics Directorate]. Other U.S. Government departments and agencies, NGOs [nongovernment organizations], and commercial partners also play a vital role in virtually all aspects of the JLEnt and function on a global scale providing comprehensive, end-to-end capabilities" (Joint Publication 4-0, *Joint Logistics*, Joint Chiefs of Staff, February 4, 2019, incorporating change 1, May 8, 2019, pp. ix–x). See also Joseph F. Dunford, Jr., "Institute for Defense Analysis Study," information memorandum to the Secretary of Defense, April 30, 2018.

[5] *Combat support* is the USAF term used to describe the collective set of functions that comprise *combat support* and *combat service support* commonly used by DoD. DoD defines *combat support* as "fire support and operational assistance provided to combat elements" and *combat service support* as "the essential capabilities, functions, activities, and tasks necessary to sustain all elements of all operating forces in theater at all levels of warfare" (Office of the Chairman of the Joint Chiefs of Staff, *DOD Dictionary of Military and Associated Terms*, Joint Staff, November 2021, p. 40). We use *combat support* (or CS) to refer to both combat support and combat service support as defined in the DoD lexicon.

Challenges for Pacific Air Forces

In response to the convergence of challenges described above, Pacific Air Forces (PACAF) has aggressively pursued initiatives to operationalize agile combat employment (ACE). ACE is an evolving operational CONEMP that seeks to improve operational resiliency when facing a capable near-peer adversary. In general, the command's ACE CONEMP is built around a more dispersed force beddown as compared with a traditional beddown, operating from clusters of interdependent hubs and spokes, and performing proactive and reactive aircraft maneuver within the cluster in an attempt to complicate adversary targeting.[6] Even with ACE, combat operations are at risk of disruption because of an adversary's sizable quiver of ballistic and cruise missiles. Logistics and sustainment support to ACE operations is further complicated by the Pacific theater's geographical characteristics. Many of the potential operating locations are separated by water and distant from supply sources located in the continental United States (CONUS). An adversary such as China, capable of creating a contested battlespace, could render runways unusable and isolate the base clusters from their sources of supply and replenishment for several days. They could also degrade communications and disable the forward forces' ability to communicate their need for additional resources.

The PACAF ACE annex to USAF's ACE CONEMP highlights that "prepositioning of equipment is essential in enabling ACE; however, assets like fuels, munitions, and spare parts require periodic replenishment," and PACAF must develop a plan for that replenishment.[7] Typically, USAF operates within a *pull system* in which the need for a resource is signaled to a supplier of that resource who delivers that it in a timely fashion to meet the need. The challenge with a pull system is that it relies on uninterrupted communication of the need and no disruptions on the ability of the system to deliver the resources, neither of which are assured in a contested environment.

Given the challenges imposed by geography and the adversary, PACAF envisions utilizing a *push system* for some CS resources that anticipates the demand for resources under varying threat conditions and levels of uncertainty and is postured to provide support without reliance on traditional lines of communication. In designing such a system, PACAF must decide (1) where to invest in additional capabilities or buffers, or both; (2) where to posture those capabilities and buffers; (3) what changes to CS tactics, techniques, and procedures (TTP) are needed, and (4) what new CS command and control (C2) authorities are needed. Making these decisions will require balancing trade-offs of the various elements of the CS system in a manner that best enables operational performance.

---

[6] PACAF, "Agile Combat Employment (ACE): PACAF Annex to Department of the Air Force Adaptive Operations in Contested Environments," Department of the Air Force, June 2020.

[7] PACAF, 2020, p. 8.

PACAF has been planning for and practicing ACE maneuver schemes that can be applied under a variety of threat conditions and environments. Headquarters PACAF/A4 asked RAND Project AIR FORCE (PAF) to assist in evaluating logistics and sustainment concepts to support the command's ACE operational CONEMP. PACAF wishes to identify how decisions it makes today regarding posture investments will enable or constrain the execution of plans being contemplated by the operational community. Rather than limiting the focus to command-owned resources, PACAF asked RAND researchers to incorporate a more holistic view of the CS enterprise, including base, theater, and global resources.

Research Approach

We conducted the research in two parts. The first part focused on identifying options for improving CS enterprise performance under various threats and uncertainties. That unclassified work considered the design of the CS enterprise, including decision authorities and the ACE operating environment, the concept of temporarily operating some part of the supply chain as a push system rather than as a pull system as conditions warrant, and investing in resource buffer stocks that could lessen the impact of unforeseen uncertainties. The second part of the work, documented in a companion report, which is not available to the general public, presents performance evaluation results of the CS enterprise in the context of ACE operations in both competition and conflict. Part of that analysis evaluates the effectiveness of some of the mitigation options suggested in this report, such as the use of buffer stocks for certain commodities and under certain conditions.[8]

The analytical results provided in the companion report provide insight about CS enterprise performance in various situations and operating conditions. However, the results of that analysis reflect enterprise performance based on a set of assumptions, the current and planned state of inventories for different commodities, and an initial CS concept of support and operation.[9] While those insights are informative, the more relevant takeaway may be the method by which those insights were gained. Using a functionally integrated analytical framework embodied in a model called the Persistent Logistics Analysis Tool and Optimization (PLATO) model, we explored the impact of varying several input parameters (e.g., location of resource inventories, adjustments to those inventories, different operational demands). The outputs of the model enabled us to identify elements of the CS enterprise that render it fragile and limit its ability to support operations. This subsequently allowed us to explore different options to mitigate poor enterprise performance. Measuring the performance of the CS enterprise in operationally relevant metrics

---

[8] Katherine C. Hastings, James A. Leftwich, Vikram Kilambi, and Ronald G. McGarvey, *Maneuvering Beyond Support: Application of the PLATO Model to Determine Sufficiency of Munitions, Fuel, and Spare Parts Required to Support a Pacific Maneuver Operation*, RAND Corporation, forthcoming, Not available to the general public.

[9] Hastings et al., forthcoming.

(e.g., sortie generation capability) is a key feature of the modeling framework presented in this report.

Organization of This Report

Chapter 2 highlights the design elements of the CS enterprise, the challenges the enterprise faces in operating in a contested environment, and strategies to mitigate those challenges. Chapter 3 provides an analysis of the mitigation options suggested in Chapter 2. In Chapter 4, we present the modeling framework used for conducting the analysis that focuses on current PACAF CS plans and posture for supporting ACE operations in the Pacific theater of operations described in the companion report, which is not available to the general public. Chapter 5 offers our observations and recommendations. We review CS enterprise decision authorities in the appendix.

Chapter 2. The Combat Support Enterprise in an ACE Operational Context

USAF's ACE CONEMP is a response to the reality that the battlespace will be contested and that the adversary will have the ability to limit freedom of movement and operations. The features of a contested environment affect not only combat operations but CS operations as well. In this chapter, we present the USAF operational approach embodied in ACE and explore the implications of the contested environment and ACE on the CS enterprise, as well as mitigation actions that could address those challenges.

ACE in the Pacific Theater

The 2018 National Defense Strategy describes a global security environment characterized by the "re-emergence of long-term, strategic competition by . . . revisionist powers," including China and Russia.[10] That environment is one in which the U.S. military advantage is challenged, and adversaries are capable of creating a battlefield that is more lethal and disruptive across multiple domains, executing operations at "increasing speed and reach."[11] These threats can limit the freedom of movement of USAF operational and support forces while inflicting damage on forces and resources. In response to this evolving threat, PACAF is reimagining its approach to employing combat forces: ACE seeks to counter the adversary's ability to strike U.S. forces and enable continued operations in a contested battlefield.

PACAF's ACE CONEMP builds on the Department of the Air Force's (DAF's) adaptive operations in contested environments (AOiCE). Fundamental to the DAF's AOiCE concept of operations (CONOP) is "proactive and reactive operational scheme of maneuver executed within threat timelines to increase survivability while generating combat power."[12] PACAF's ACE CONEMP builds on those fundamentals and integrates the command's concept of cluster basing, which focuses on operating from a "distributed network of hub and spoke locations to maintain operational momentum."[13] This fundamental maneuver is described in the ACE CONEMP as *agility*, one of the four elements that define ACE, and the CONEMP points out that logistics

[10] DoD, *Summary of the 2018 National Defense Strategy of the United States of America: Sharpening the American Military's Competitive Edge*, 2018, p. 2.

[11] DoD, 2018, p. 3.

[12] PACAF, 2020, p. 2.

[13] PACAF, 2020, p. 2.

enables the agility needed to operate in a contested environment.[14] Another element of the ACE CONEMP is *posture*. The CONEMP explicitly states that posture actions include the prepositioning of supplies and equipment to enable combat operations.[15]

The final two elements of the PACAF ACE CONEMP are *joint all-domain command and control* (JADC2) and *protection*.[16] JADC2 and protection, along with agility and posture, have implications for logistics and, more broadly, the CS enterprise. Where should logistics and CS resources be prepositioned, and in what quantities, to enable agile combat operations in a contested environment? How will C2 of those resources be executed in a conflict? How should they be protected against adversary actions that can degrade logistics capabilities, and what mitigation actions can be applied in establishing the posture to ameliorate the impact of potential adversary actions?

Fundamentals of the USAF Combat Support Enterprise

The CS enterprise consists of 24 functional areas that, through combined efforts, enable USAF forces to generate air combat power projection, as well as support forces on the ground.[17] Those areas closely tied to combat power projection include fuels, munitions, supply, maintenance, and distribution. Other CS functions, such as civil engineering, services, security forces, and medical, to name just a few, focus more on support to deployed forces at a forward operating locations.

The functional areas within the CS enterprise utilize a large variety of resources, including fuels, munitions, general purpose vehicles, large construction vehicles and equipment, tents, aircraft and vehicle spare parts, and many others. Each resource has different characteristics with respect to how easily it can be moved, its costs, its availability, how frequently it is needed, and its importance to different parts of the USAF mission. Those characteristics factor into posture considerations. For example, items that are difficult to transport or needed very early in a conflict are often prepositioned at or near the intended point of use.

We limited our analysis on those functions and resources most closely tied to combat power projection. Thus, we considered resources such as munitions, fuels, and spare parts. We also included repair capabilities and transportation in our analysis because these functional areas enable the resupply of munitions, fuels, and spare parts for combat force generation.

---

[14] PACAF, 2020, p. 3.

[15] PACAF, 2020, p. 3.

[16] PACAF, 2020, p. 3.

[17] Air Force Doctrine Publication 4-0, *Combat Support*, LeMay Center for Doctrine, January 5, 2020.

General Design and Characteristics of the Combat Support Enterprise

USAF policies for the CS enterprise operations explicitly describe assumptions, responsibilities, and authorities for design, management, and execution decisions in both competition and conflict. As we highlighted earlier, the CS enterprise has moved to centralized management of many commodities in an attempt to seek efficiencies. Key resources and functions, such as munitions, spare parts, fuels, and strategic transportation, are managed by single organizations from a global perspective. The benefits and drawbacks of centralized management are well documented.[18]

When firms move to a global management strategy, they tend to operate with a centralized organizational structure in which decisions, information, and resources are directed to tightly coupled subsidiaries.[19] Centralized management allows firms to realize efficiencies in their supply chain and allows companies more flexibility to reallocate resources in response to long-term disruptions.[20] Despite these benefits, one of the main drawbacks of centralized management is the inability to respond rapidly to fluctuations in demand or disruptions caused by a rapidly changing environment.[21]

Beyond centralized management, USAF policy bases the management of contingency materiel on a set of principles that may not be realistic in contested settings, including (1) "the materiel management system must be structured to provide uninterrupted support for both in-place and deploying forces," (2) "day-to-day operations must mirror contingency operations to the greatest extent possible to minimize disruption and training disconnects," and (3) "base self-sufficiency and resupply management responsiveness must be maximized."[22] Planning for conflict also includes a series of assumptions that may not be realistic in contested settings. For examples, "[US]AF doctrine is to establish immediately premium transportation based air routes for eligible Class IX (a) and Class VII (x) assets from point of use to repair node and retrograde to point of use to achieve consistent resupply within 72 hours,"[23] as well as "from the beginning of a contingency until normal materiel management operations are resumed, materiel management support will consist of POS [peacetime operating stock], RSPs [readiness spares package], follow-on RSPs and responsive 'Express Delivery Service' of high priority items."[24]

---

[18] See David Simchi-Levi, *Operations Rules: Delivering Customer Value Through Flexible Operations*, MIT Press, 2013.

[19] Jahangir Karimi and Benn R. Konsynski, "Globalization and Information Management Strategies," *Journal of Management Information Systems*, Vol. 7, No. 4, Spring 1991, p. 11.

[20] George S. Yip, "Global Strategy . . . in a World of Nations?" *Sloan Management Review*, Vol. 31, No. 1, Fall 1989, p. 33.

[21] Yip, 1989, p. 34.

[22] Air Force Instruction (AFI) 23-101, *Materiel Management*, Department of the Air Force, July 8, 2021, p. 118.

[23] AFI 23-101, 2021, p. 104.

[24] AFI 23-101, 2021, p. 119.

The fundamental design of, and assumptions about, USAF logistics operations are challenged by contested environments.

Combat Support Operations in a Contested Environment

Adversary actions in a contested environment can take many forms and affect CS operations in numerous ways. Attacks on information systems and networks could result in degradation of communications between forward operating locations where resources are needed and the decisionmakers and materiel managers who are making and executing allocation decisions. Attacks on runways could constrain the ability of the distribution system to send resources where they are needed. Damage to inventory control points and warehouses could result in the loss of supplies and equipment that are necessary to conduct combat operations. The primary challenge is the uncertainty of knowing when, where, or how the adversary might try to limit USAF operations.

Prior PAF research addressed uncertainty in conflict in a report called *Coupling Logistics to Operations to Meet Uncertainty and the Threat (CLOUT).* The CLOUT report distinguishes between two types of uncertainty in conflict: (1) statistical uncertainty linked to "variability observed in repeatable phenomena" and (2) state-of-the-world uncertainty, or "uncertainties about phenomena that are not repeatable, not observed or observable, or both."[25] USAF deals regularly with statistical uncertainty and has processes and policies in place to handle it. An example is in forecasting demand for spare parts to support the annual flying hour program. In this case, the system used for forecasting demand for spare parts takes an eight-quarter moving average of historical demand, applies some smoothing techniques using different distributions, and produces a forecast.[26] Contested environments, however, are fraught with state-of-the-world uncertainty. Operations could be of higher intensity than originally planned. Resources and infrastructure could be destroyed or damaged, leading to resource shortfalls or restricted lines of communication. Exactly how conflict with an adversary will unfold cannot be predicted with 100-percent accuracy.

The CLOUT analysis highlights that a central consideration for the performance of the logistics system is *response time*, which can include time to repair, process and handling times, order-and-ship times, time to deliver or transport, and even communication times.[27] For our analysis, we use the observe-orient-decide-act (OODA) loop to structure activities related to

---

[25] I. K. Cohen, John B. Abell, and Thomas F. Lippiatt, *Coupling Logistics to Operations to Meet Uncertainty and the Threat (CLOUT): An Overview*, RAND Corporation, R-3979-AF, 1991, p. v.

[26] Air Force Materiel Command Manual 23-101, Volume 5, *Equipment Specialist Data and Reports*, November 17, 2016, certified current December 15, 2021.

[27] Cohen, Abell, and Lippiatt, 1991, p. 6.

logistics response time.[28] The *observe-orient-decide* elements associated with resource allocation in conflict are tied to decision authorities within the CS enterprise and are subject to delays resulting from adversary actions affecting the communications network's ability to convey demands in a timely manner. The *act* component of the OODA loop becomes a function of how quickly resources can be delivered to the point of need, which subsequently becomes a function of the location of the needed resources and the availability of transportation to get it to its point of need. The fundamental goal of the enterprise is to keep the logistics OODA loop operating inside the operational OODA loop, in other words, making sure the ability to resupply critical resources is accomplished within the time they are needed, to avoid disruptions in the operational mission.

Snyder et al. (2021) also discuss uncertainties associated with operating in a contested environment and conclude that CS enterprise performance will be affected by any number of variables that cannot be predicted, although the possibility of degraded communications is more likely than not.[29] *Command and Control of U.S. Air Force Combat Support in a High-End Fight* focuses explicitly on the structure, systems, and mechanisms for C2 of logistics resources in a communications-degraded environment. Similar to the operational context of our analysis, it also focuses on logistics support to a maneuver force with an enterprise-wide view of CS stakeholders.[30]

The 2021 report breaks down the challenges associated with operating in a communications-degraded environment; two challenges, in particular, are relevant to the motivation for our analysis. First, communications-degraded environments demand that the CS enterprise have "the ability, at least for brief periods, to distribute command and control of combat support."[31] The authors note that logisticians at a forward operating location may not have the means of signaling the need for additional resources and may need to make localized decisions regarding the utilization of available resources. Second, communications-degraded environments demand "the

---

[28] The OODA loop concept is built on the notion that a person or force can gain a strategic or tactical advantage against an adversary by adapting to changes in the environment and making decisions faster than the adversary does, thus creating confusion in the mind of the adversary who is executing their plan on outdated information. Our application of the OODA loop concept is more basic and mechanical than the original concept developed by Colonel John Boyd (USAF). We use the more basic model simply as an approach to think about the elements of logistics response time and the need for the logistics OODA loop to operate inside the operational OODA loop not to confuse or disorient operational planners but rather to enable operators to accelerate their own OODA loop. Boyd never formally published his work; for a more complete elaboration on it, see Robert Coram, *Boyd: The Fighter Pilot Who Changed the Art of War*, Back Bay Books/Little, Brown and Company, Hachette Book Group, 2002, pp. 334–344.

[29] See Don Snyder, Kristin F. Lynch, Colby Peyton Steiner, John G. Drew, Myron Hura, Miriam E. Marlier, and Theo Milonopoulos, *Command and Control of U.S. Air Force Combat Support in a High-End Fight*, RAND Corporation, RR-A316-1, 2021.

[30] Snyder et al., 2021, p. 3.

[31] Snyder et al., 2021, p. 9.

ability to operate temporarily with limited situational awareness."[32] The implication here is that logisticians in the command center will not know what resources are available at the forward operating location and may need to predict need and proactively send additional resources.

The centralized management approach, coupled with the type of logistics system operated by USAF, was also addressed by Snyder et al. (2021), who noted that, as a means of achieving efficiency in the CS enterprise, USAF operates a *pull logistics system* in which resupply actions are triggered by a demand from a using entity. (This is in contrast to a *push logistics system*, which will be explored further in the next chapter, in which resupply actions are proactive and based on estimated demand.) The authors go on to highlight that pull logistics systems "are more fragile in the face of communications and data loss."[33] The challenges this presents in communications-degraded environments can be more easily overcome when decision authorities are more decentralized and resources can be allocated and reallocated by logistics commanders who are more informed about the nature of the operations than when enterprise-level resource managers are attempting to make decisions with less situational awareness.

Conducting operations in a contested environment has implications for CS decision authority that span strategic resource planning, planning for conflict, and executing in conflict. Decisions made during the planning, programming, budgeting, and execution system process affect the CS enterprise in two ways: the ability to effectively demonstrate ACE during competition as a means of deterring conflict and the way in which resources are postured across the enterprise to support combat operations should conflict occur. Decisions about where to posture resources affect the *act* element of the logistics OODA loop because resources postured closer to the point of need reduce the response time.

During conflict, the decision authorities are directly tied to the ability of the component major command (C-MAJCOM) staff, operating in a wartime structure as the Air Force forces (AFFOR) staff in support of the commander of AFFOR and the joint force air component commander, to support air operations. Therefore, these authorities align with the *observe-orient-decide* elements of the logistics OODA loop. Presumably, the AFFOR staff will have better insight on the demand for resources and the status of the inventories *within* the theater than some other entity operating remotely at a centralized resource control point. Similarly, logisticians at forward operating locations would have even greater situational awareness relative to their base clusters.

Contested operations force USAF to rethink several questions related to decision authorities:

- Who has authority to make decisions about commodities and activities?
- How do decision authorities change from competition to conflict?
- How does ACE and a high-threat environment affect decision authorities?

---

[32] Snyder et al., 2021, p. 10.

[33] Snyder et al., 2021, p. 26.

- Under what conditions would decision authorities change, and for how long?

Where decision authorities rest in competition affects the posture of the theater for conflict. In some instances, decisions regarding the locations and quantities of resources in the theater are made at the enterprise-level and may be made with limited insight to the operation plans (OPLANs) of the theater. How those decisions are made in competition versus conflict can affect the logistics OODA loop, which, as we highlighted, is an important measure of the CS enterprise. How the decision authorities might change in a high-threat environment is also closely linked to the logistics OODA loop and plans that the theater might need to consider in a communications-degraded environment. Those plans will likely need to focus on adjustments to the decision authorities and the boundaries of those adjustments, which are addressed by the fourth question.

Decisions made during competition about quantities of resources required, how many or much to buy, where to allocate the resources, how to posture them globally, and how to manage them in support of competition activities (such as ACE exercises) can affect PACAF's ability to effectively demonstrate ACE operations as a deterrent and be prepared to engage in conflict on short notice. Decisions made during execution regarding requirements, allocation and reallocation priorities, and methods of distribution can affect the agility and responsiveness of the logistics system in a dynamic environment.

Given the current design of and policies directing the conduct of the CS enterprise, as well as the challenges introduced by contested operations, an approach to mitigating such challenges might seem unfeasible. All the same, we offer, and subsequently analyze, some mitigation strategies to address risks imposed by the uncertainties of conducting CS operations in a contested environment.

Approaches to Risk Mitigation

We focused on two primary mitigation strategies that USAF could consider to address risks to CS operations in a contested environment. The strategies generally fall into CS enterprise *design* actions and *posture* actions. Enterprise-wide design actions address fundamentals of USAF logistics operations and decision authorities, while posture actions focus on the location and quantities of resources. Both Snyder et al. (2021) and Cohen, Abell, and Lippiatt (1991) offer thoughts about mitigation strategies that could be explored to address a communications-degraded environment and the uncertainties experienced during conflict. *Analysis of Global Management of Air Force War Reserve Materiel to Support Operations in Contested and Degraded Environments* addresses posturing strategies of the CS enterprise in contested environments.[34] We introduce those mitigation strategies here and analyze their value in subsequent chapters.

Addressing Design Flaws

Snyder et al. (2021) observe that "the ability to adjust command and control of logistics when under persistent multi-domain attack is impeded by the lack of doctrine, policy, planning, and procedures for distributed command and control and for push logistics,"[35] and they suggest that USAF needs to "establish process and planning factors to generate the demand signal."[36] The authors go on to point out that USAF may need to consider a blended push-pull system or perhaps even temporarily a pure push system in which resources are sent to the anticipated point of need based on historical consumption factors and predicted usage. We examine a push system as a mitigation strategy in more detail in Chapter 3.

In our analysis, we explored other considerations that should be incorporated in these process and planning factors. In the context of a push system, a high-end fight abounds with uncertainties. Will transportation resources be available or airfields accessible? Over what time intervals will the push system need to operate? Which resources need to be pushed?

Even with a primary, alternate, contingency, and emergency (PACE) plan, the issue of decision authorities in a contested environment must be addressed. Decision authorities in competition, where demands are fairly predictable and communications are available, are generally sufficient to enable the logistics OODA loop to respond with just-in-time delivery of parts and other resources. In conflict, when the operational community is likely seeking to shorten its operational OODA loop, the logistics OODA loop is at risk of lengthening. Degraded communications could prevent high-echelon decisionmakers from observing the need for resources at forward locations, thereby delaying the decision to reallocate resources. Delays in the *act* element of the logistics OODA loop would likely be affected by the availability of transportation assets to move resources and the distances they would need to travel given the vast Pacific theater of operations. Some of these vulnerabilities can be overcome by posture decisions, which we discuss later; however, keeping the logistics OODA loop inside the operational OODA loop in conflict will likely require a multifaceted approach that addresses all elements of the OODA loop construct. Addressing decision authorities regarding how to best utilize on-hand resources must shift to more forward-based entities that have immediate visibility into resource inventories, and required operations is one place to start.

---

[34] Kristin F. Lynch, Anthony DeCicco, Bart E. Bennett, John G. Drew, Amanda Kadlec, Vikram Kilambi, Kurt Klein, James A. Leftwich, Miriam E. Marlier, Ronald G. McGarvey, Patrick Mills, Theo Milonopoulos, Robert S. Tripp, and Anna Jean Wirth, *Analysis of Global Management of Air Force War Reserve Materiel to Support Operations in Contested and Degraded Environments*, RAND Corporation, RR-3081-AF, 2021.

[35] Snyder et al., 2021, p. 30.

[36] Snyder et al., 2021, p. 41.

There are actions USAF can take today to enhance its existing processes. To begin, USAF can acknowledge in doctrine, policy, instructions, and manuals that the adversary may attack (kinetically or non-kinetically) and communications may be lost, thus forcing a different approach to USAF processes.[37] That different approach should be outlined in TTP for each functional area. Because each functional area is different—some have one global manager, some have many, some are organic to USAF, some are not—these TTP should be function-specific and still provide the same type of information as they do now.

Next, TTP should contain PACE plans for each functional area. Units at the base regularly report information (status, activities, resource requests) through communications channels, such as a computer system update, an automatically generated report, or a report produced with Microsoft Office products (Excel, Word). Each functional area should have PACE plans to convey that unit-level information to the information users clearly defined in TTP or other instructions.

Each functional TTP should also contain the battle rhythm or schedule for reporting information for that functional area. Because the criticality of receiving information about each functional area differs during conflict, each battle rhythm should be based on when that functional area's information is needed. For example, munitions management, airfield operations, and force support are all critical functions; however, the information needed by users may occur on different timelines. Munitions expenditures and on-hand munitions inventory may need to be reported when aircraft return from a mission or in preparation for the next launch. Runway status and radar capabilities may need to be reported more often, such as hourly, so other units are aware that the forward location is open and operating. Force support may need to supply casualty reporting daily or every 12 hours. Whatever the schedule of information is, it should be established and documented before conflict begins.

The TTP should also identify the trigger for when to begin operating within the functional battle rhythm. The trigger could be an order from the combatant command (CCMD) or a request from the AFFOR staff in advance of combat operations. Or it could be reactive and begin as soon as units begin to fly combat operations. Whatever the trigger is, it should be defined in the TTP for each functional area.

If communications capabilities are denied or degraded to the point where the PACE plan is ineffective, the TTP should outline how each functional area could move to push logistics, as needed. Not every functional capability will need to push resources when communications are lost, but some will. The logistics support concept and timeline for those decisions should be clearly defined in each functional area's TTP. For example, once a battle rhythm has begun, the

---

[37] There may be other reasons to move to another approach, such as natural disaster (hurricane, earthquake, flood, tornado) or civil unrest.

identified information user (this could be the global resource manager or the AFFOR staff) will receive information updates according to the schedule established in the TTP.

For highly critical resources, the information user may expect updates several times within a 24-hour period. If the user does not receive updates within the expected time frame, they may begin to prepare to move to a push system. They could gather last-known update information and expected resource usage or status. If the unit is unable to provide information at the next scheduled update, the information user can begin to calculate the push package. Then, if a third update is missed, depending on expected inventory and criticality, the information user could initiate the push package. Again, a push package is not necessary for every functional area, and the timeline will vary. We offer the above timelines as notional examples of the kinds of activities and timelines that should be specified in the TTP for each functional area.

Once the TTP define the processes and the timelines, the information users (whether that is a global manager, a centralized control point, or the AFFOR staff) should practice developing a push package. They need access to the most recent information on the system that will contain that information. They need access to information about expected activity from, perhaps, the air tasking order (ATO).

Forward operating units may develop mission plans on their own if they do not receive a daily ATO. The PACAF air operations center (AOC) is developing processes to continue operations in a communications-degraded environment using a baseline ATO and guidance packages. These processes will allow forward units to continue operating even if they do not have a current ATO. Information users should understand the processes that forward units will use to develop their mission plans so they (information users) can use a similar process to calculate expected resource usage.

In addition, information users should communicate with other global managers to share information. A situation may occur in which one information user does not receive an update, but another user may have received a more current update (because of reporting frequency) or information through another communications link. For example, the battle rhythm for a munitions update could be scheduled for every six hours, whereas a fuels update may only be received every 12 hours. If communications are denied and the munitions staff have a more current update on the number and types of missions flown (because they received a munitions update), that information should be shared with other functional areas (such as the CCMD's joint petroleum office [JPO]) so those other functional areas can calculate a push package using the most current information from the forward units.

On the unit side, once a unit is unable to provide and confirm that an update has been sent to an information user, authorities for unit-assigned and collocated assets should transfer to that unit. With numerous functional communities likely represented at a forward location, that location should have a lead authority, such as the base operating support integrator, the lead squadron commander, or another who can make decisions about priorities, missions, and safety.

That unit lead authority's decision will inform unit-level functions, such as how to allocate and prioritize resources within a functional area.

Addressing Posture Issues

Operating a hybrid push-pull system as part of the general design of CS enterprise operations in a contested environment will likely require a shift in the CS enterprise posture, including the types, quantities, and locations of CS resources required to enable ACE operations in a contested environment. Additionally, a push system will often include a reallocation of resources, lateral resupply, and responsive transportation to create a network of buffer stocks to mitigate the possibility of incorrect forecasting given the state-of-the-world uncertainties that are prevalent during a war.[38] These are the "uncertainties about phenomena that are not repeatable, not observed or observable, or both" mentioned earlier.[39] The CLOUT research discussed two key principles on which the concept for dealing with uncertainty is based. Both were relevant when the report was published in 1991, and one is still relevant today.

The first principle, which is arguably not applicable today, was that "logistics operations be based on demands as they become known in real time and as they are predicted more reliably over *very short horizons* [original emphasis]."[40] The nature of operating in a contested environment influenced by adversary actions that can degrade communications makes the likelihood of demand *being known* in real time less likely. That said, the ability to *predict demands* over short time horizons will serve as the basis for estimating the investment required to enable a push system. The need to consider resource inventories to establish uncertainty buffer stocks requires a more detailed consideration of the types of resources suitable for a push system that optimally balances resource availability and responsiveness and that is economically feasible.

The second principle is that logistics must be tightly coupled with operations and that the effectiveness of the logistics systems should be measured in operational terms.[41] We find this principle to be valid still today. Similar to what is suggested in the CLOUT analysis, we consider resource allocation decisions and lateral resupply as viable means to achieve operational sortie production requirements. In our analysis, we differentiate *resource allocation* as being the determination of where to direct resources flowing into theater from CONUS or centralized supply points and *lateral resupply* as being the "reallocation of resources across units."[42] In the latter case, we view this in the context of units operating within the same ACE cluster.

---

[38] Cohen, Abell, and Lippiatt, 1991, p. 20.

[39] Cohen, Abell, and Lippiatt, 1991, p. 6.

[40] Cohen, Abell, and Lippiatt, 1991, p. 16.

[41] Cohen, Abell, and Lippiatt, 1991, p. 17.

[42] Cohen, Abell, and Lippiatt, 1991, p. 17.

Furthermore, the CLOUT analysis highlights the challenges with a "buyout" strategy. A *buyout strategy* suggests simply acquiring sufficient resources needed to reduce the risks of parts not being available because of errors due to statistical uncertainty. Cohen, Abell, and Lippiatt (1991) highlights that even in benign wartime scenarios, a buyout strategy may be insufficient.[43] In a contested environment, a buyout strategy would be likely to fail as well because of the high degree of state-of-the-world uncertainties. For our analysis, we recognize that a buyout strategy might not be feasible economically, but we explore the tradespace in terms of costs, responsiveness, and the impact on aircraft availability relative to operational demands. Because a full buyout strategy is often infeasible, a strategy centered on buffer stocks would be more likely. Thus, USAF would need to address the question of where to position these buffer resources.

In considering global management of war reserve materiel (WRM), Lynch et al. (2021) discuss the dynamics of WRM management and the trade-offs that must be considered, highlighting that joint policy suggests that decisionmakers should seek "the greatest practicable flexibility to respond to a spectrum of regional contingencies while reducing the burden on the global transportation network."[44] The authors introduce a decision tree framework for a global posture of capabilities that considers ease of movement, costs, and timing of need.[45] Using that framework as a foundation, we develop a similar framework to consider what resources might be candidates to consider as part of a push system. We present this framework in Chapter 3 and demonstrate its use by applying it to spare parts, including the additional characteristic of criticality. These characteristics, however, become factors when considering how to posture the network to enable a responsive logistics OODA loop.

Some resources are cheap and readily available, making them easier to acquire in sufficient quantities without a large investment. Others are expensive, making it costly to attempt to acquire additional stocks. There are resources that might be expensive but difficult to move, making them candidates for prepositioning closer to the intended point of use if the objective is to minimize airlift requirements. Bucketing resources according to these various characteristics is necessary to assess the cost and benefit of different strategies and approaches for posturing the CS enterprise to deal with uncertainty and threats.

Other Mitigation Strategy Considerations

Although addressing design and posture flaws is an important step toward improving CS enterprise operations in contested environments, other institutional challenges persist. The CS enterprise is a collection of stakeholders dependent on the decisions and activities of other

---

[43] Cohen, Abell, and Lippiatt, 1991, p. 9.

[44] Chairman of the Joint Chiefs of Staff (CJCS) Instruction 4310.01F, *Logistics Planning Guidance for Pre-Positioned War Reserve Materiel*, Joint Chiefs of Staff, August 29, 2022, p. A-1.

[45] Lynch et al., 2021, pp. 30–32.

enterprise stakeholders all working to deliver support to the warfighter. Those stakeholders exist at every level within the enterprise, each with varying degrees of control over the decisions and resource allocations that affect posturing for the high-end fight in competition and executing the high-end fight in conflict.

While CS processes are well defined and well understood in planning and execution during both competition and conflict, management of many resources is divided across several organizations within and outside USAF. It can be challenging to get resources to the right places when needed because of the coordination that must occur across these organizations. In a contested, ACE-like environment in particular, where aircraft and ground support for those aircraft are moving and the adversary is attacking, it will become more challenging to continue some processes as they are currently defined. Combat units may not be able to communicate their needs to those with allocation authorities, and allocating authorities may not have visibility into inventory levels in the theater. An enterprise-level organization, that holds the authority to reallocate resources from one theater to another, may be slow to process the request for reallocation, or the bureaucratic process for approving the request may result in delays in the logistics response time.

Additionally, current policies and processes can inhibit the performance and responsiveness of the CS enterprise in conflict settings. The enterprise relies on accurate forecasting of resource requirements in peacetime and conflict, timely transmission of requests for parts or resources, timely delivery by assured transportation, and allocation decisions based on enterprise-level agencies assuming complete situational awareness. Under normal conditions, the enterprise should operate efficiently and deliver effective results. In a conflict setting, particularly a contested ACE conflict setting, many of these underlying elements will not hold.

Similarly, programming and posturing decisions are based on long-standing planning factors contained in the War Mobilization Plan–Volume 5 (WMP-5).[46] Are the planning factors contained in the latest version of the WMP-5 consistent with the emerging and evolving CONEMPs that rely on force maneuver and distributed operations as a means of improving operational resiliency?

The reality is that even with adjustments to decision authorities in conflict, there are disconnects in CS enterprise processes during competition and in what resources the theater controls in conflict. Some of these challenges can be mitigated by a posture strategy that places buffer resources in the theater closer to the point of use and more in the control of the warfighting commander. As noted in the CLOUT analysis, a *pure* buyout strategy would not be feasible. The question then becomes, what are the cost and benefits of a *partial* buyout strategy? Although we address both of these strategies in Chapter 3, the latter is the focus of more in-depth analysis in the companion report, which is not available to the general public.

---

[46] DAF, *War and Mobilization Plan 2011*, Vol. 5: *Basic Planning Factors and Data*, October 2010.

Chapter 3. Assessing Mitigation Strategies Related to Combat Support Enterprise Design Flaws and Posture Modifications

Earlier, we highlighted that USAF must consider mechanisms to mitigate potential uncertainties that may occur during competition and conflict. Prior research suggests that a *complete* "buyout" strategy would likely be infeasible because of the significant costs that must be incurred.[47] Cost is still a significant limiting factor today. More recent research suggests the need for a temporary push logistics system or hybrid push-pull logistics system to contend with the likelihood of degraded communications and that buffer stocks may be necessary to mitigate the uncertainty with forecasting future demands based on historical demands.[48] Other recent research suggests elements of the tradespace that must be considered when determining where and what to preposition, recognizing the required balance between costs, risk, ease of transportation, and timing of need.[49] We incorporate those prior research observations in this chapter while exploring two potential mitigation strategies developed in response to likely uncertainties faced by the CS enterprise.

Assessing a Push System as a Mitigation Strategy

Operating a hybrid push-pull system, while a new approach for USAF, is not a novel solution in industry. In his book *Operations Rules*, David Simchi-Levi describes a push-pull system as one in which some elements of the supply chain operate on a push-based model, and other elements operate on a pull-based model.[50] Simchi-Levi writes about matching the supply chain strategy to the type of products needed and provides a framework for selecting the appropriate strategy, push versus pull, to employ based on several factors. His framework focuses on demand uncertainty, lead time, and factors in economies of scale.[51] In the case of PACAF considering a strategy for spare parts, we provide a similar framework and expand the factors that should be considered in the design and operation of a push-pull system.

In establishing our notional push-pull system, we first determined which resources were appropriate to push and then where those resources should be pushed from. We considered five characteristics for key commodities: frequency of demand, criticality for mission success, cost,

---

[47] Cohen, Abell, and Lippiatt, 1991.

[48] Snyder et al., 2021.

[49] Lynch et al., 2021.

[50] Simchi-Levi, 2013, p. 37.

[51] Simchi-Levi, 2013, pp. 40–45.

weight, and availability. The first two characteristics (frequency and criticality) determine whether a part should be included in the push system or remain in the traditional pull system. For those resources that we included in the push system, we used cost to determine whether an item should be pushed from the CONUS or within the theater, and we used the last two characteristics (weight and availability) to determine the storage location of the resource (a centralized storage facility or multiple dispersed facilities or operating locations). The framework for this approach is depicted in Figure 3.1.

For each step, a value threshold for the associated characteristics must be applied to appropriately categorize the resource. For example, when considering frequency of demand as the first filter, one could establish a threshold such that for any resource required on average less than once every five days, that item will remain in the pull bucket, while any resource needed more frequently than that would become part of a push system. The thresholds need not be arbitrary. Five days for frequency, for example, could be based on plans for communication recovery or airfield repair that would result in a return to normalcy within that period.

Figure 3.1. Push Versus Pull Analysis Framework

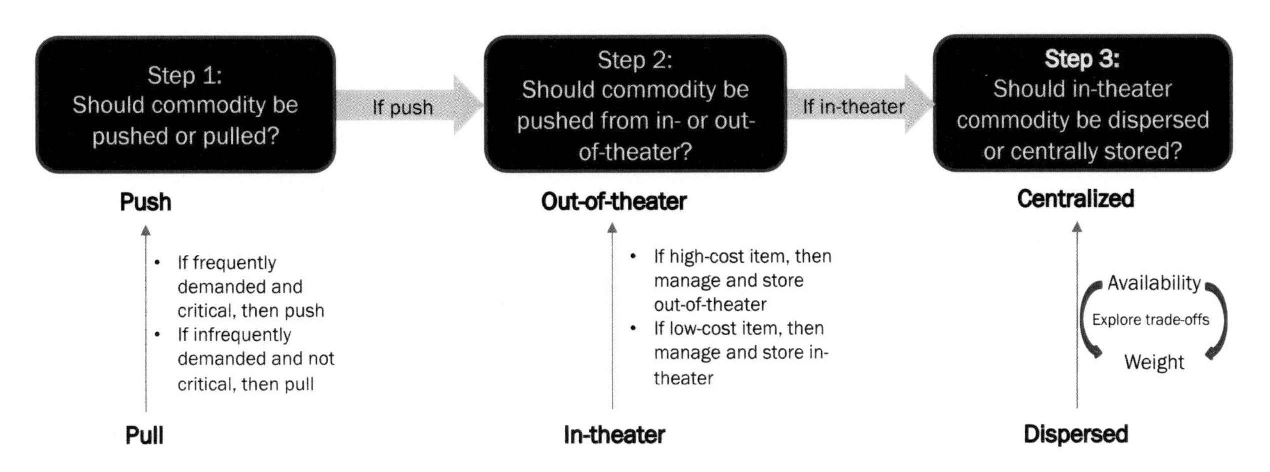

Although this framework is designed to evaluate any resource or commodity in the CS enterprise, we will illustrate our approach by applying it to spare parts for the F-15C. Thus, when we refer to "parts" in the process described below, these are interchangeable with any resource or commodity in the CS enterprise. To assemble a list of parts to evaluate, we use the repairable parts included in RSPs. We evaluate each part independently in our analysis.

Step 1: Frequency and Criticality

Frequency of demand is important in order to establish a rhythm for the push system and to establish a reasonable limit on the number of commodities in the push system. The idea being that if a resource is required infrequently and the push system is intended to operate intermittently when communications or access to forward locations are degraded, the resource could likely remain in the pull system and the request for resources transmitted when the degradation was remediated.

To determine how frequently a part is needed, we determine the expected number of days of supply provided by the RSP inventory. The expected demand is calculated from failure rate data and the operations tempo (OPTEMPO) of our notional flying scenario as defined by the number of aircraft, sortie rate, and average sortie duration (ASD) in hours:

$$expected\ demand\ per\ day = failure\ rate * number\ of\ aircraft * sortie\ rate * ASD \quad (3.1)$$

We then use this expected daily demand to estimate the number of days of supply (for a given part) provided by the quantity (Qty) provided in the RSP as follows:

$$supply = \frac{RSP\ Qty}{expected\ demand\ per\ day} \quad (3.2)$$

Parts with an on-hand supply that fall under the specified frequency threshold are considered high-frequency demand items. For example, if a five-day frequency threshold has been set, a part with only three days of on-hand supply provided by the RSP would be considered a high-frequency part.

Additionally, some parts may be deemed critical to flying operations (or, in the context of the general framework, resources that are critical to providing support for the forces) that are not considered to be high-frequency demand items based on the identified threshold. In some cases, it may be of interest to include these critical parts (resources) in a push system to mitigate against uncertainty of demand that may arise in conflict, particularly in communications-degraded situations where demand cannot be conveyed, to ensure continuity of operations. It is up to the planner whether to consider the criticality of parts to the mission (as well as frequency) and to include critical parts in the push system.

Critical parts (resources) can be incorporated in two ways. In one approach, all critical parts could be included in the push system, although this could result in a large number of parts (resources) within the push system. Alternatively, a secondary, more relaxed (and, thus, inclusive) frequency threshold could be set for those critical parts (resources) that do not meet the standard frequency threshold. For example, an initial 10-day threshold may be set for parts within an RSP, such that any part having less than a 10-day supply in the RSP would be included in the push system. Spares of mission impaired capability awaiting parts (MICAP) might then be subject to additional consideration. In the first approach, all MICAP would also be included in the push system regardless of the RSP inventory. In the alternative approach, a secondary threshold for MICAP might be set, for example, a 20-day threshold such that any MICAP with less than 20 days of supply provided by the RSP are also included in the push system. Those MICAP with more than 20 days of on-hand supply in the RSP would remain in the pull system with all non-MICAP that have more than ten days of supply provided by the RSP.

The threshold values that are used should reflect the shorter of the anticipated duration of operations (which can be rather short under the ACE CONEMP) and the estimated time that a line of communication might be interrupted by adversary action. Because these values are uncertain, a range of values should be considered; the resultant impact on the number of parts in the push system should be quantified and presented to decisionmakers to aid in the threshold value selection process.

Step 2: Item Cost

Cost of the commodity is included in the framework to assist in determining from where it should be pushed. In this step, the assumption is that some commodities are low-density items because they are extraordinarily expensive, and USAF limits the quantities it purchases. It is more likely that such items would be centrally stored. For items identified in step 1 as candidates for a push system, those with high cost would be considered high-demand/low-density items, and USAF may want to carefully manage their distribution and push them from a warehouse in CONUS. Lower-cost items, which might be more abundant, could be pushed from a theater warehouse. As with frequency of demand, a threshold for cost would need to be established. For example, a \$100,000 threshold could be set such that any part in the push system with a unit cost less than \$100,000 may be prepositioned within the theater, while parts in the push system exceeding this threshold are sourced from CONUS. In general, determination of this threshold would likely be an iterative process in which a threshold is set, then parts identified as being pushed from in-theater would be evaluated on whether they are high-demand/low-density items, and the threshold updated as desired.

In the case of our spares analysis, we pulled cost information for each part from Federal Logistics Data (FED LOG). FED LOG allows users to search a list of National Stock Numbers (NSNs) to find cost, weight, and other information to inform logistics decisions.[52]

Step 3: Availability and Weight

After determining that a part should be pushed and that it should be pushed from inside the theater, we then determine whether the part should be stored in a central location in the theater or further prepositioned and distributed in the theater. This final step requires balancing two factors, availability and weight. For example, how available is the part at the various locations across the theater? Will airlift even be available to move the part and, if so, how much airlift is expected to be available?

---

[52] We accessed NSN data in July 2021 from Defense Logistics Agency (DLA), "FED LOG–Federal Logistics Data," database, undated.

In our analysis, DLA provided monthly snapshots of the on-hand serviceable quantity for parts owned by USAF and DLA, for each mission design series (MDS).[53] We used the average monthly serviceable quantities for PACAF- and CONUS-based warehouses to determine the average number of days of supply available in the system.[54] A threshold for on-hand supply can then be established, such that any part in the push system that is to be prepositioned in-theater with fewer days of on-hand supply than the threshold would be centralized in theater and those parts with more days of on-hand supply would be dispersed within theater.

The weight of parts is also considered to account for the potential desire to reduce the intra-theater airlift burden. If concerns about lift availability are prevalent, the desire might be to further distribute the parts to reduce or nearly eliminate response time. In the case of completely eliminating response time by positioning parts at their expected point of use, the parts would essentially become prepositioned buffer stock added to the RSP. Similar to cost, we pulled information on part weight from FED LOG.

The question in applying the third step of the framework centers on the objective of deciding where to locate pushed items. Is the intent to reduce airlift requirements (optimize lift) or accelerate logistics response time (optimize response time)? The answer to this question then drives the thresholds set for availability and weight.

Results of the Push-Pull Analysis for the F-15C

We now apply our framework to the F-15C, using a 24-aircraft squadron and its associated RSP. We do this to highlight how USAF might consider which resources might be included in a push system if converting to a hybrid push-pull system. Current policy calls for fighter aircraft RSPs to be designed for 30 days of operations at the sortie rate and sortie duration outlined in the WMP-5.[55] However, current RSP composition does not provide 30 days of supply for all parts, as shown in Figure 3.2. This analysis focuses on repairable parts within the RSP for which failure rate data were available. We identified 131 repairable parts (by National Item Identification Number) with an expendability-recoverability-repairability category code of XD2, indicating that they are expendable depot recoverable items managed under USAF's Recoverable Assembly Management System. Failure data were available for 127 of these parts. As part of our analysis, we examined a range of ASDs as a sensitivity analysis against uncertainty, recognizing that the WMP-5 planning factors might not represent what is currently envisioned by PACAF for ACE sortie rates and durations.

---

[53] DLA, Analytics Center of Excellence, "DLA Distribution Receipts," monthly data feed to RAND, January 2021.

[54] The estimated days of on-hand supply is determined based on warehouse inventory and expected daily demand, which is determined by the total expected flying hours.

[55] DAF, 2010, p. K-1. Information confirmed with Richard A. Moore, Headquarters Air Force Materiel Command/A9A, email discussion about RSPs with the authors, September 9, 2021.

Figure 3.2 presents the percentage of parts in each RSP for which the authorized levels are not sufficient to support 15 and 30 days of supply, as a function of the assumed ASD of 24 aircraft flying with a sortie rate of 1.0 (i.e., 24 sorties per day). Here, we see that 20 percent of the parts do not have an RSP inventory to support 15 days of low OPTEMPO flying operations (4-hour ASD), and 50 percent of parts cannot support 15 days of high OPTEMPO operations (10-hour ASD). Similarly, 44 percent of parts do not have the RSP inventory to support low OPTEMPO operations for 30 days, and 70 percent of parts cannot support operations for this long under high OPTEMPO conditions.

Figure 3.2. Number of F-15C RSP Parts with Less Than 15 and 30 Days of Supply, by ASD

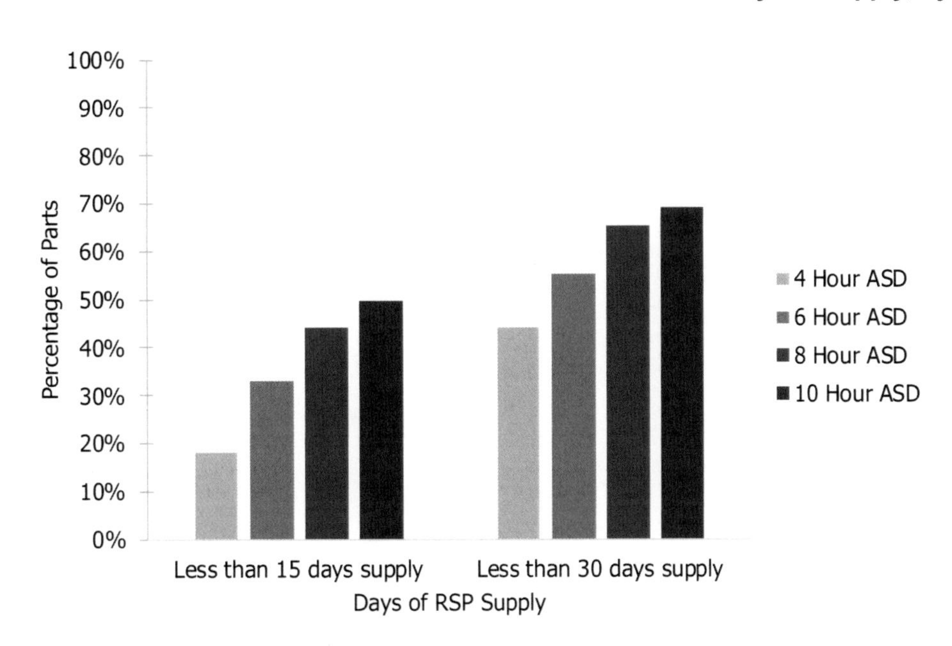

NOTE: Our calculations include only the 127 repairable parts for which failure rate data were available.

Step 1: Determining Parts in Push Versus Pull System

To begin, appropriate threshold values must be selected for the number of days of supply for MICAP and for non-MICAP. (Note that 70 of the 127 parts included in our analysis are considered MICAP.) As discussed previously, this step involves professional military judgment to estimate the envisioned duration of operations and the duration that an adversary might be capable of interrupting the lines of communication between an operating location and the CS network. The intensity of operations, represented by the ASD, imparts another level of uncertainty into this threshold determination. Figure 3.3 presents an example, in which a variety of threshold values, for both the standard frequency and critical (MICAP) frequency, are considered. Observe that at the largest threshold values (30-day standard frequency threshold and 60-day threshold for MICAP), the number of parts in the push system ranges from 71 to 97, for an ASD of four and ten hours, respectively, out of 127 total parts in the RSP considered in this analysis.[56] As the thresholds are decreased, the number of parts in the push system decreases because the thresholds are more restrictive in terms of which parts should be pushed.

Figure 3.3. Number of F-15C Parts in Push System

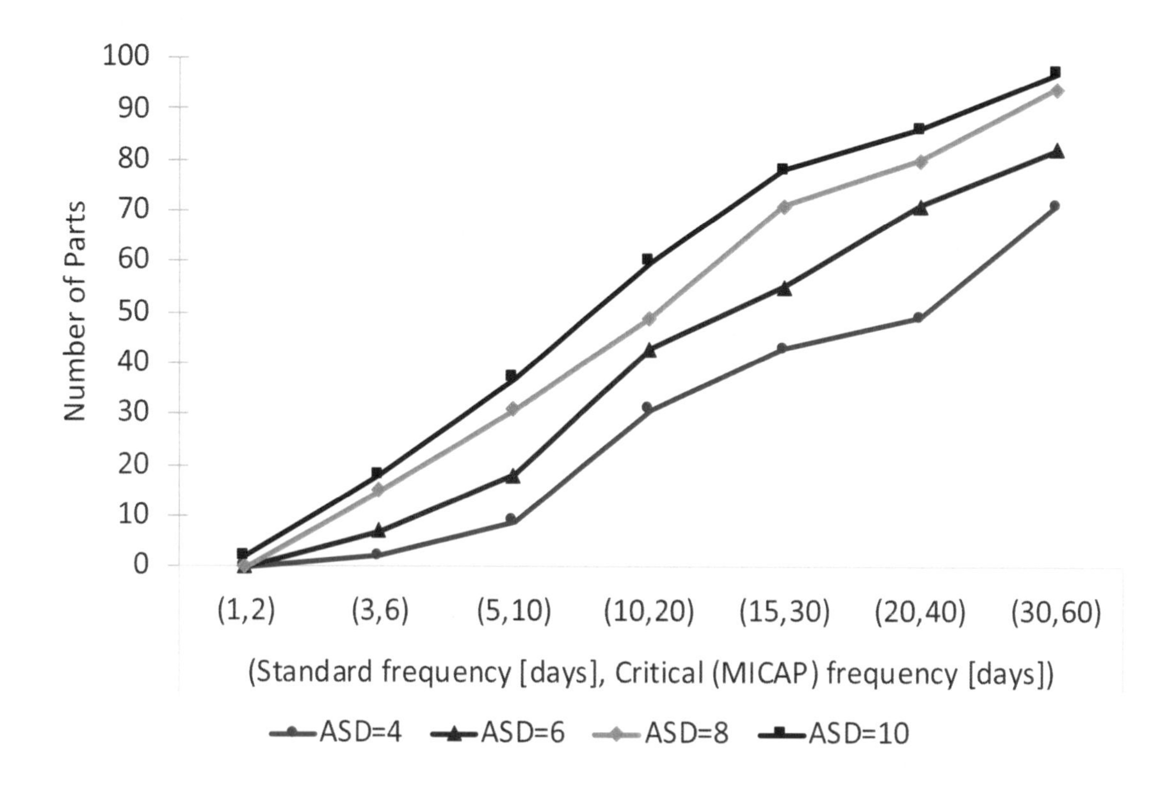

Suppose that, based on this information, PACAF selects a threshold of ten days of non-MICAP supply and 20 days of MICAP supply. For these thresholds, Figure 3.3 suggests that between 31 and 60 parts might be included in the push system, depending on the expected ASD. Figure 3.4 shows the breakdown of parts between the pull and push systems for the indicated thresholds as a function of ASD. For those parts that are candidates for a push system, we must now determine whether they should be pushed from CONUS or prepositioned in theater.

---

[56] Recall that the PACAF F-15C RSPs contain a total of 131 reparable spare parts (we do not consider consumables), and for four of these, failure rate data were either unavailable or identified to be zero (i.e., zero failures per flying hour). Thus, these four parts were omitted from our analysis.

Figure 3.4. Number of F-15C Parts in Push and Pull Systems for a 10-Day Standard Frequency Threshold and 20-Day MICAP Frequency Threshold

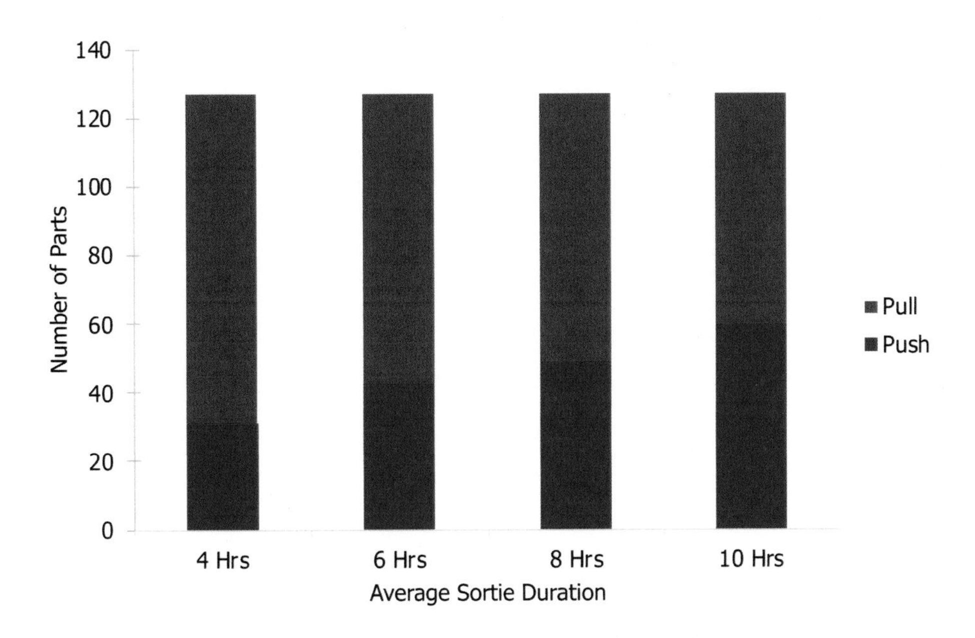

NOTE: Hrs = hours.

Step 2: Determining In-Theater Versus Out-of-Theater Storage for Pushed Parts

In step 2, we consider the parts that have been identified in step 1 as candidates for a push system to determine whether they should be pushed from CONUS or prepositioned in theater. For demonstration purposes, we consider two cost thresholds—$25,000 and $100,000—such that any part whose cost is less than the cost threshold is prepositioned within theater and any part whose unit cost is more than the cost threshold is stored out of theater in CONUS. As illustrated in Figure 3.5, we see that, of the 31 spare parts that fall into the push system under a four-hour ASD, 11 of them would be prepositioned in theater when a $25,000 cost threshold is set, whereas 23 of them would be prepositioned within theater if a $100,000 cost threshold is set. In general, the number of spares in the push system that may be prepositioned in theater ranges from 11 (for a 4-hour ASD and $25,000 cost threshold) to 48 (for a 10-hour ASD and $100,000 cost threshold).

Figure 3.5. Number of F-15C Parts in Push System, by Storage Location

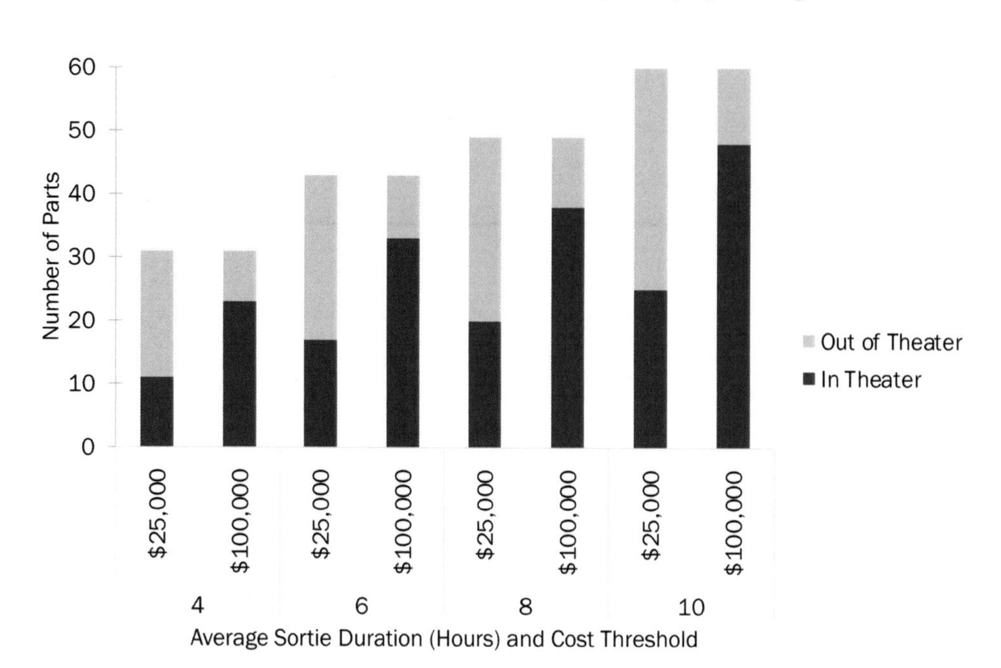

Step 3: Considering Centralized Storage or Dispersal in Theater for Pushed Parts

The final step is to assess whether parts pushed from an in-theater location should be centrally stored or further prepositioned and distributed within theater. Here, two characteristics—weight and availability (in USAF- and DLA-owned warehouses in CONUS and PACAF)—are considered and a threshold established for each. (Recall that these thresholds should be based on a trade-off analysis between transportation requirements and response time. The thresholds set here are for demonstrative purposes only.)

We demonstrate this step for the set of parts corresponding to the $25,000 cost threshold for each ASD (see Figure 3.5). We consider a notional weight threshold of 10 pounds and a notional availability threshold of 20 days of on-hand supply so that any part whose weight *and* warehouse availability are less than those thresholds would be centrally stored within theater and any part that exceeds both of these thresholds would be dispersed within theater. For those parts that exceed only one of the thresholds, planners would need to decide how such parts should be stored within theater.

The graph in Figure 3.6 plots the availability and weight of all parts that have been identified as being stored in-theater as part of a push system for the frequency, criticality, and cost thresholds identified in Steps 1 and 2 and for the four ASDs considered. Furthermore, we have presented availability as a function of each of the four ASDs considered so that the availability/weight trade-off can be considered simultaneously in the context of competition (shorter ASDs) and conflict (longer ASDs) because prepositioning must occur during competition but support conflict. However, this results in multiple data points for most of the parts. For the 11 parts identified as candidates for in-theater storage at a 4-hour ASD (see

Figure 3.5), there are four distinct data points (one for each of the four ASDs considered). Similarly, there are three data points for each of the six additional parts to be stored in theater when ASD increases from four to six hours (one for each of the ASDs of six hours and above); two data points for the three additional parts that are to be stored in theater should the ASD increase from six to eight hours (one for an 8-hour ASD and one for the 10-hour ASD); and one data point for the additional five parts that are stored in theater when the ASD increases from eight to ten hours. Note that weight is a static parameter here, so the set of corresponding data points occur along a horizontal line.

Figure 3.6. F-15C Parts in Push System Stored Centrally in Theater Versus Dispersed in Theater

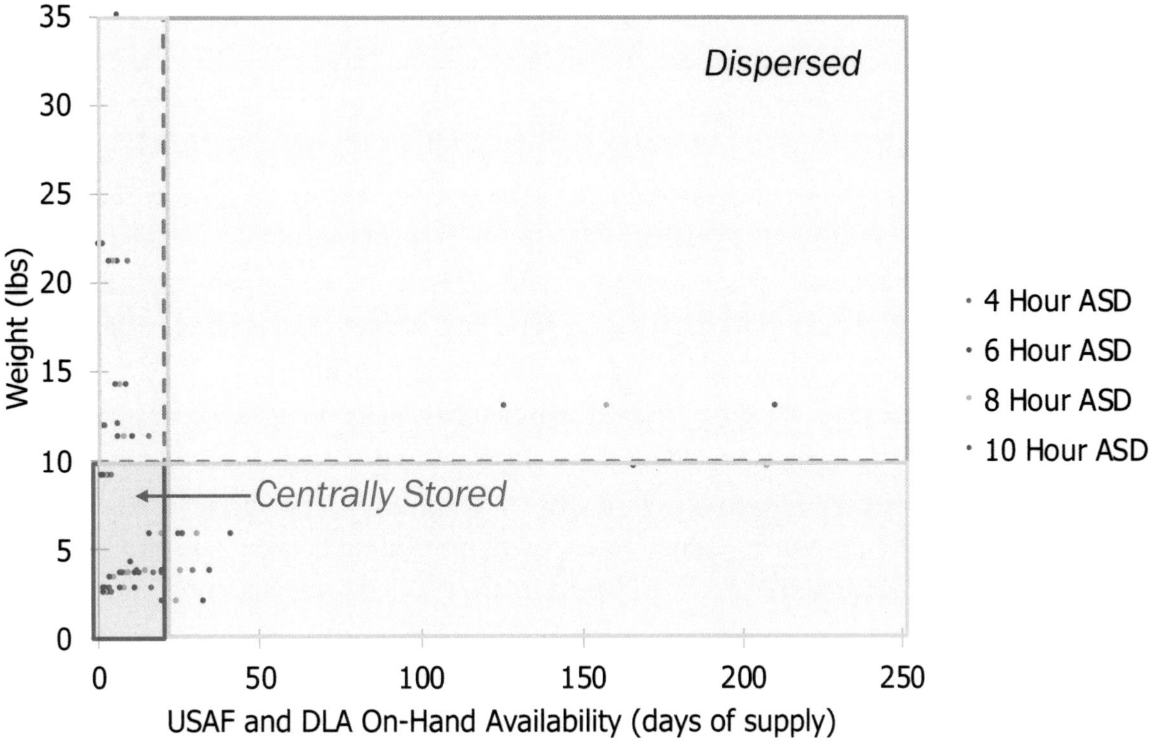

Dispersed

Centrally Stored

Weight (lbs)

USAF and DLA On-Hand Availability (days of supply)

· 4 Hour ASD
· 6 Hour ASD
· 8 Hour ASD
· 10 Hour ASD

We have also plotted the 10-pound weight threshold (horizontal dotted line) and 20-day warehouse availability threshold (vertical dotted line) to provide the reader with a sense of the number of parts that fall above or below each threshold. The orange shaded area in the lower-left area indicates the set of parts that should be centrally stored within theater based on our notional thresholds, and the blue shaded area in the upper-right area indicates those parts that should be dispersed within theater. The other two gray-shaded areas represent situations in which parts exceed only one of the two thresholds and independent decisions may be necessary to determine the prepositioning of each part. In situations where some data points corresponding to a given part fall within an area of the graph that suggests either a centralized or dispersed posture, while other data points fall within the gray, ambiguous area, it likely makes sense to follow the posture suggested by the associated data points. (Note that because data points corresponding to the same part occur along a horizontal line, we are guaranteed that data points corresponding to a given part could only fall into either the centralized or dispersed storage category, but not both, so there will be no confusion.) In situations where all data points corresponding to a given part fall within a gray area of the chart, additional consideration may be necessary, or the analyst may wish to adjust thresholds to remove or reduce ambiguity.

Of the 25 parts that have been identified for storage within theater, this particular aspect of the analysis identifies nine parts that should be stored centrally (the weight and availability thresholds suggest central storage regardless of ASD), one part that should be stored in a more dispersed fashion, two parts that should likely be stored centrally (depending on the ASD, data points are sometimes assigned to central storage and sometimes the assignment is ambiguous because both parts exceed the warehouse availability threshold, suggesting dispersed storage, but neither exceed the weight threshold, suggesting central storage), and 12 parts for which a posture cannot easily be determined (their associated data points fall within the gray area for all ASDs considered, making their assignment ambiguous). Additionally, one part had no warehouse inventory indicated in the dataset available at the time of this analysis, so its posture could not be determined. Given that the data points associated with 12 parts do not definitively determine posture assignment via these thresholds, an analyst might wish to examine each part more closely or consider alternative thresholds to make final determinations.

In the case of spare parts, the airlift savings would likely be minimal; however, the cost of parts would likely be high, which would tilt the scale toward more centralized storage. If other commodities were being considered as part of a temporary push system, the airlift savings of pushing more parts from forward locations as opposed to a central location in theater could produce more significant results.

Mitigation Strategies to Increase Operational Resiliency

As reflected in Figure 3.2, numerous parts in the F-15C RSP do not have sufficient inventory to support 15- or 30-day operations (without resupply), even in the case of short ASDs. Here, we consider two mitigation strategies to increase operational capability via increased aircraft availability: a complete buyout strategy and a buffering strategy. The previously referenced CLOUT analysis addressed both, noting that a *complete buyout* (i.e., purchasing the necessary inventory to bring all parts in the RSP up to a 15- or 30-day supply) would likely be infeasible because of cost. That said, we consider the cost of a complete buyout and conduct a buffering analysis to explore the dynamics of a partial buyout and an example of the uncertainty one might mitigate against.

We demonstrate our mitigation strategies using the same F-15C spare parts considered in the push-pull framework analysis. We first identify the cost associated with a *complete buyout* strategy, that is, the cost to increase RSP inventory to ensure sufficient parts to support operations across the *desired* conflict length and ASD.[57] Then, recognizing that the extremely large cost of such a strategy likely renders it infeasible, we vary the number of days of supply that could be bought out, show the associated reduction in cost, and discuss the differences in assumed risk.

In this analysis, we treat the ASD value as the uncertain parameter. In Chapter 2, we noted that forecasting and programming for spare parts are based on planned sortie rates and expected sortie durations found in the WMP-5. The WMP-5 breaks these factors out by theater of operation and weapon system type, which are expected to reflect the combatant commander's (CCDR's) plan. There are two challenges with this. First, the WMP-5 was last published in 2010, and the planning factors may not reflect the CCDR's current strategy for prosecuting a conflict. Second, the plan may simply be a plan, and as CLOUT points out, war is fraught with state-of-the-world uncertainties.

Buyout Strategy: Increase RSP Quantities

We calculated the spare part requirement assuming that the RSP will support 24 aircraft, each flying one sortie per day. Eight RSP requirements were calculated, one for each combination of conflict length (15 days or 30 days) and ASD (4, 6, 8, or 10 flying hours), using the formula shown in Equation 3.3. (If the calculation resulted in a requirement for a fractional number of parts, we rounded up that number to the next integer.)

$$part\ requirement = \lceil failure\ rate * number\ aircraft * sortie\ rate * ASD * conflict\ length \rceil \quad (3.3)$$

We next identified the investment necessary to buy out the RSP requirement by subtracting the existing RSP authorized quantity for each part from the requirement obtained from Equation 3.3, then multiplying this value by the part's unit purchase price. The results of these calculations appear in Figure 3.7, which presents the buyout cost to increase the existing RSP quantity to the appropriate size (for either a 15- or 30-day kit) based on each combination of conflict length and ASD. We see that the cost to increase a single F-15C RSP to a full 15-day kit ranges from $2 million (for a 4-hour ASD) to $18 million (for a 10-hour ASD), and the cost to increase it to a 30-day kit ranges from $12 million to $53 million. These costs are per squadron and in addition to the $100 million cost of the repairable items already authorized in these RSPs.

[57] We consider a complete buyout with respect to necessary inventories to support conflict operations because RSPs are intended to be used during conflict rather than in competition.

Figure 3.7. Buyout Cost to Increase Existing PACAF F-15C RSP to a 15- or 30-Day Kit

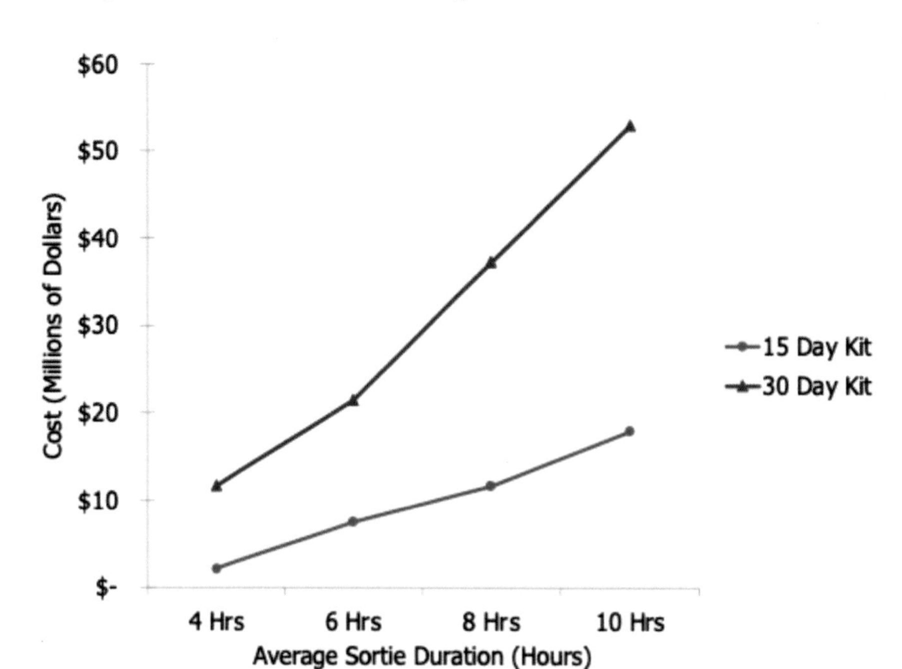

The cost to increase RSP levels via a buyout strategy would likely be prohibitive, particularly when these kits must be designed to support long ASDs. As an alternative, we next consider an alternative purchasing strategy in which a subset of parts is buffered against, that is, an additional inventory for a subset of parts (of interest) would be procured. Although this additional inventory would not increase the kit inventory to a full 15- or 30-day capability, it would mitigate against uncertainty in the CS enterprise's ability to sustain operations during conflict.

Buffering for Five Days of Requirements

Similar to the push-pull analysis presented earlier, we identified the set of parts against which we wish to buffer by setting thresholds for the RSP inventory of non-MICAP and MICAP. As before, for demonstration purposes, we set a standard frequency threshold of 10 days and a MICAP threshold of 20 days. For this set of parts, ranging from 31 parts for a 4-hour ASD to 60 parts for a 10-hour ASD, we considered procuring an additional five days of (buffer) stocks that would be stored in theater, likely at a cluster hub. (Note that the parts to be buffered and the quantities that must be purchased to support five days of operations depends on the ASD.) Figure 3.8 reflects the cost to procure these buffer stocks based on the assumed sortie duration compared with the full buyout cost. The cost to procure five days' supply of these parts is approximately $6 million, which is greater than the cost to increase the RSPs to 15 days' worth of supply (by approximately $3.5 million) but less than the cost to increase to a 30-day RSP, for the same ASD. For ASDs of six hours or more, we see that the five-day buffer stock is cheaper to procure that a complete buyout. In fact, the cost to procure a five-day buffer stock for 10-hour

ASDs is approximately 33 percent less ($12 million versus $18 million) than the cost of a complete buyout for a 15-day RSP.

Figure 3.8. Cost to Procure 5-Day Buffer Stock Versus Complete Buyout

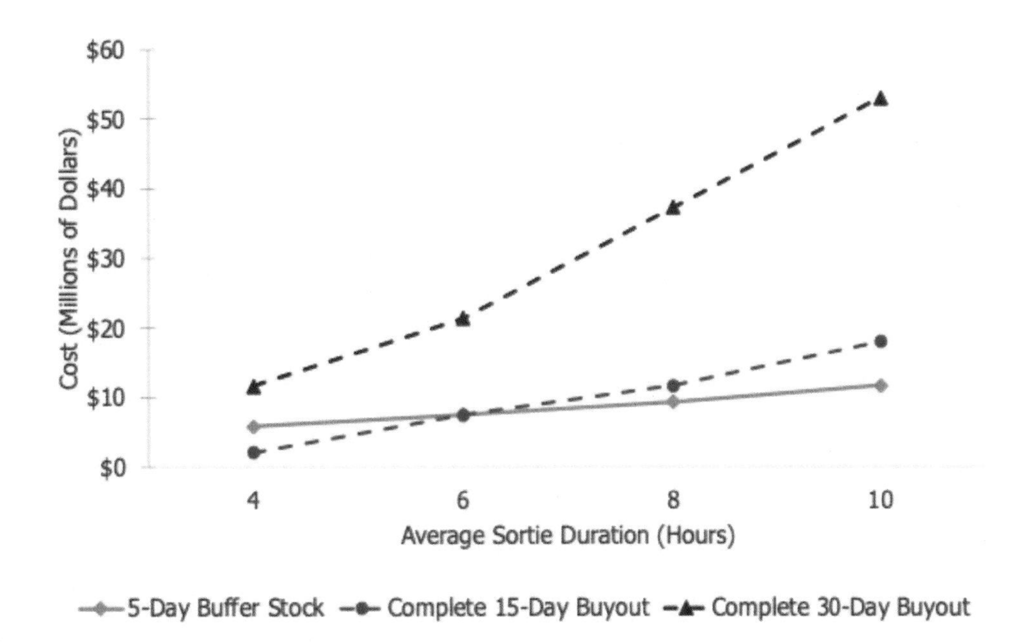

Results of Buffer Strategy Analysis

These buyout and buffer strategies present two approaches to address the state-of-the-world uncertainties USAF might expect to encounter in combat operations with a near-peer. With a complete buyout strategy, USAF purchases a relatively wide breadth of parts (between 23 and 63 different parts for 15-day conflict operations and between 56 and 88 different parts for 30-day conflict operations, depending on ASD). With the buffer stock strategy, on the other hand, USAF purchases a narrower breadth of parts (between 31 and 60 different parts, depending on ASD). This is because the buffer stock strategy focuses its purchases on those parts for which the current RSP authorizations are insufficient to satisfy the expected initial 10 days of demand for non-MICAP or the expected initial 20 days of demand for MICAP. Said differently, the buffer stock strategy concentrates its risk on non-MICAP for which the current RSP quantity authorizations would be expected to support more than 10 days of operations but fewer than 15 or 30 days. The buffer stock strategy does not accept any increased risk for MICAP in the event of a 15-day conflict, but it does accept increased risk for MICAP for which the current RSP quantity authorizations would be expected to support more than 20 days of operations but fewer than 30 days. Note that the complete buyout strategy for a 15-day conflict, as examined here, actually accepts increased risk for MICAP whose current RSP authorizations would be expected to support more than 15 days of operations but fewer than 20 days.

Moreover, for non-MICAP that are included in this buffer stock strategy (all of which are necessarily included in the 15-day buyout strategy for equal ASD value), the buffer stock strategy always results in purchasing a quantity that is less than or equal to the 15-day buyout strategy. This is because the different strategies are accepting risk in different ways. The buyout strategy assumes that the parts in the RSP should be sufficient to meet expected demands without any connectivity to the CS enterprise for the entire planned conflict length. Note that this buyout strategy implicitly accepts risk by increasing its initial transportation requirement for deployment. For example, the 15-day buyout/10-hour ASD strategy increases the RSP footprint from 24 stons to 27 stons. Figure 3.9 presents a scatterplot showing the cost and weight of the existing RSP, along with each of the options considered for RSP buyout.

Figure 3.9. Additional Weight of Full RSP Buyout Strategy

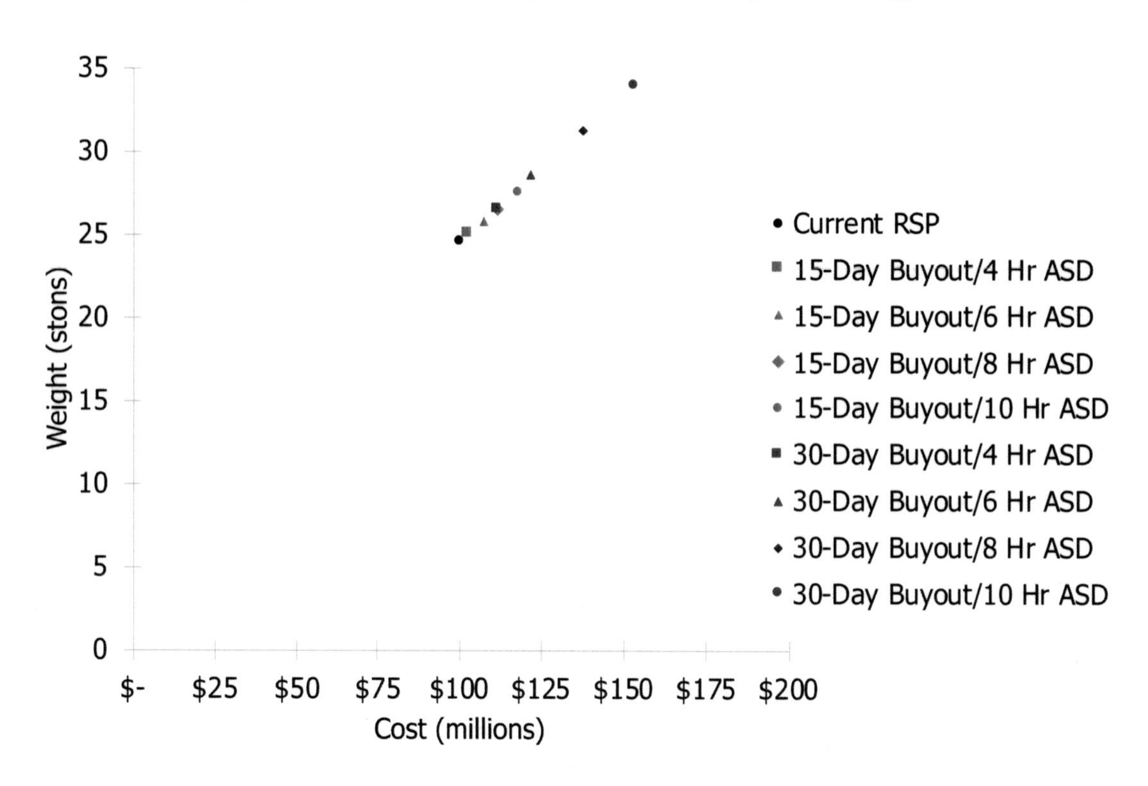

NOTE: We have used the term *ston* to refer to a U.S. ton (2,000 lb), as DoD does, to make the distinction between it and a long ton or British ton, which is 2,240 lb.

Conversely, the buffer stock strategy assumes that inter-theater lift is going to be available at some point within the conflict and trades off this increased reliance on the CS enterprise during conflict against a reduced transportation requirement during the squadron's initial deployment. For the comparable 10-hour ASD buffer stock, the initial RSP footprint does not increase from 24 stons, but it is now assumed that the 5-day buffer stock, constituting 2.1 additional stons and corresponding to less than one C-130 sortie, can be pushed to the unit using intra-theater lift starting no later than day 10 of the conflict. Figure 3.10 presents a similar scatterplot showing the cost and weight of the existing RSP, along with each of the buffer stock options that we considered in our analysis.

Figure 3.10. Cost and Weight of a 5-day Buffer Strategy

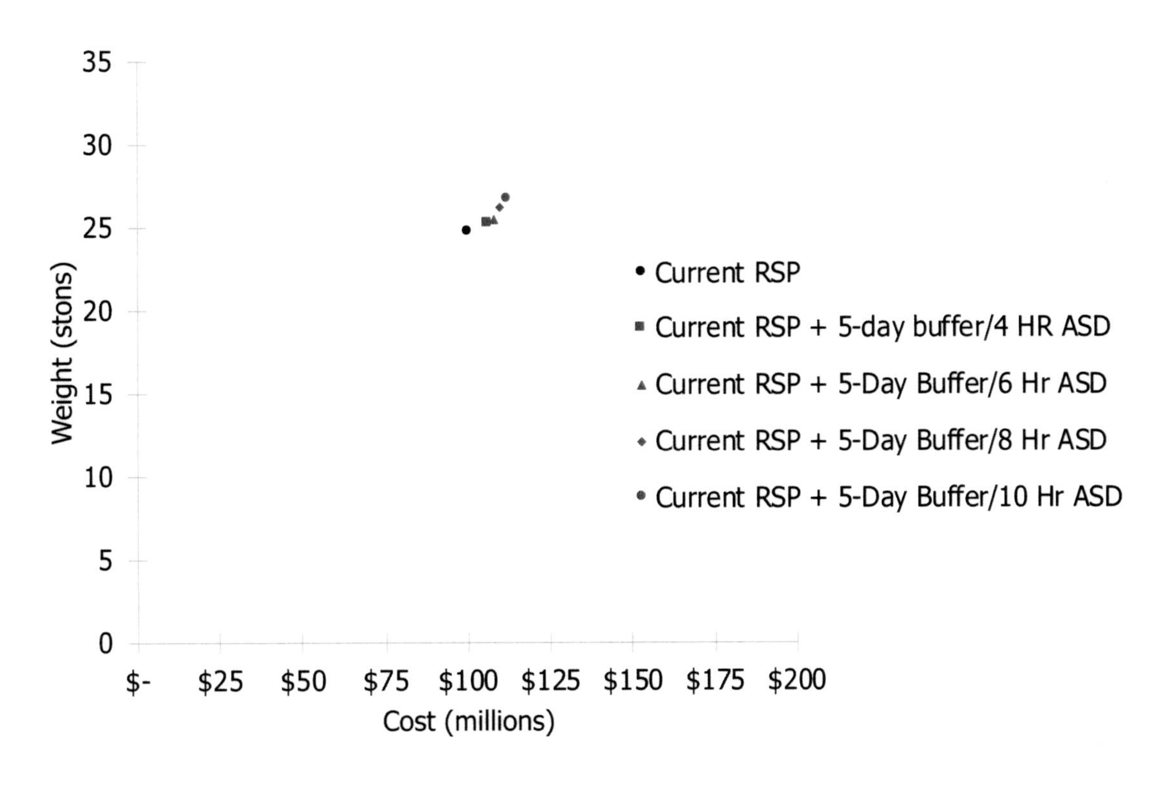

NOTE: We have used the term *ston* to refer to a U.S. ton (2,000 lb), as DoD does, to make the distinction between it and a long ton or British ton, which is 2,240 lb.

Key Takeaways

- Current RSP quantity authorizations are based on 2010 WMP-5 planning factors that do not reflect current operational plans. Thus, RSPs are insufficient to support current or future conflict operations with a near-peer.
- A complete buyout strategy to increase RSPs to full 15- or 30-day capability would likely be cost-prohibitive.
- Partial buyouts and buffer stocks can offer a more cost-effective solution to mitigating against risks caused by state-of-the-world uncertainty. Our analyses in the companion report, which is not available to the general public, demonstrate the operational capability that can be gained through a buffering strategy.[58]

[58] Hastings et al., forthcoming.

Chapter 4. Modeling Methodology to Assess the Effectiveness of Different Posture Options

Overview of the Methodology for Determining ACE Support Postures

Given the disconnects between the existing CS enterprise and the capabilities necessary for ACE, USAF needs an analytical framework to quantify, in operationally relevant terms, the impacts of resource and capability shortfalls on emerging CONEMPs. Both materiel (e.g., munitions, fuels) and non-materiel (e.g., minimum information required to maintain situational awareness) resources and capabilities must be integrated across echelons (e.g., globally, in theater, squadron level) to provide the CS required by ACE. An understanding of how these factors contribute, as a unified system, to the desired mission-generation outputs (e.g., sortie production) is necessary to make effective recommendations for future CS resourcing and C2 investments.

A Mathematical Modeling Framework for Assessing Combat Support to ACE Operations

To integrate these factors within a unified modeling framework, RAND researchers developed a tool known as the PLATO model. Previous RAND research on CS informed this model development; particularly important references include the CLOUT initiative from the early 1990s and the Agile Combat Support (ACS) body of work from the early 2000s.[59] The ACS research envisioned the creation of separate models for each important resource (e.g., munitions, fuels, shelters, spare parts, engines) along with an integrating optimization model that identified a mix of support options capable of satisfying sortie generation requirements. While many of the considerations in the ACS framework remain relevant today, ACE operational concepts exist within a landscape that has changed significantly over the past 20 years. In response to a contested operating environment, with significant risk of adversary attack to both operating and support locations, ACE envisions force maneuver within a cluster of bases in which aircraft can change their operating location frequently, for varying time intervals, while no longer requiring flying and maintenance unit integrity at a single site. Moreover, as the JLEnt has shifted its focus from effectiveness to efficiency, CS resource control has been increasingly centralized, requiring the analysis to move beyond the base (or even theater) level and capture global-level resourcing and C2.[60] Non-materiel considerations, such as C2 information flows,

[59] See Cohen, Abell, and Lippiatt, 1991; and Robert S. Tripp, Lionel A. Galway, Timothy Ramey, Mahyar Amouzegar, and Eric Peltz, *Supporting Expeditionary Aerospace Forces: A Concept for Evolving the Agile Combat Support/Mobility System of the Future*, RAND Corporation, MR-1179-AF, 2000.

[60] Tripp et al., 2021.

potential adversary actions, and authorities (e.g., push versus pull logistics), need to be incorporated into modeling efforts to permit a fuller exploration of alternative operational and logistical CONOPs.

PLATO models the CS system using differential equations, similar to approaches that have been used for deterministic inventory models.[61] The discretization as difference equations with fixed timesteps helps delineate between CONOP-level strategies and operational detail with some loss of fidelity. Resource utilization levels are assumed to vary with the planned sortie requirements in a deterministic fashion (e.g., at a given timestep, six F-22 sorties are planned, each of which consumes a fixed amount of fuel and each of which expends a fixed amount of munitions). Although the model as presented in this report assumes that this resource utilization occurs in a deterministic fashion, the overall modeling framework could allow analysts to relax this assumption and incorporate stochastic considerations (e.g., random failures of spare parts as a function of the flying hours accrued).

To help visualize the model, Figure 4.1 depicts the relationships between inputs, model processing, and outputs that are used by PLATO. The figure also demonstrates how the model processes align with stages of the logistics OODA loop.

---

[61] D. K. Bhattacharya, "On Multi-Item Inventory," *European Journal of Operational Research*, Vol. 162, No. 3, May 2005; Veronika Novotná, "Numerical Solution of the Inventory Balance Delay Differential Equation," *International Journal of Engineering Business Management*, Vol. 7, 2015.

Figure 4.1. PLATO Model Framework

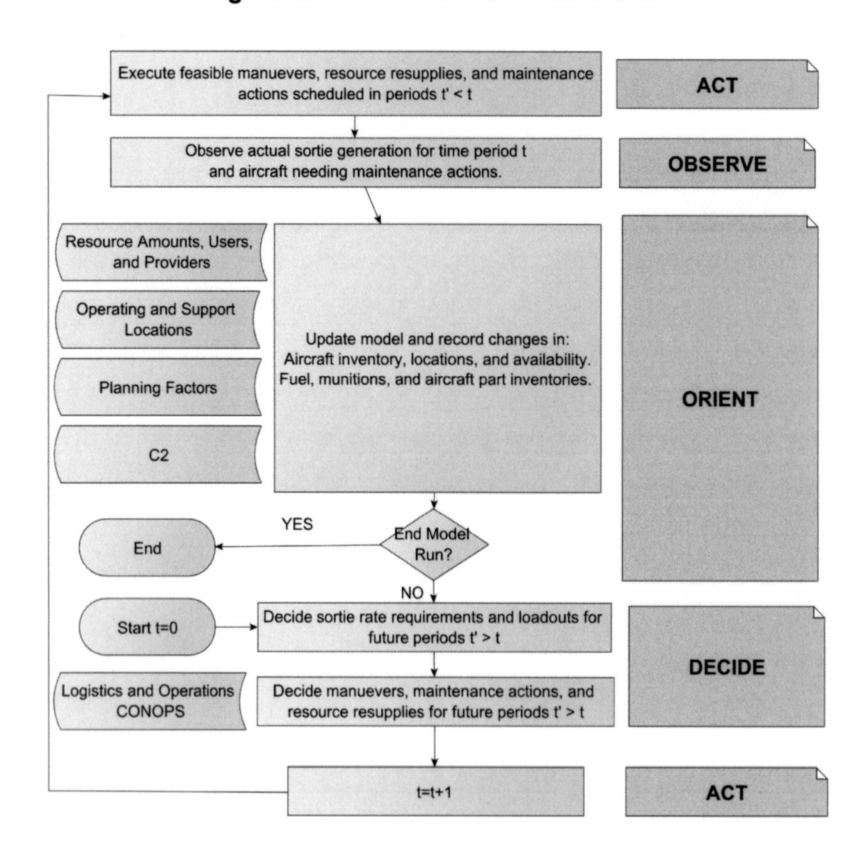

NOTE: *t* denotes the time variable in the model.

Inputs for Assessing Combat Support to ACE Operations

The PLATO model is intended to permit the examination of a variety of operational and logistics CONOPs. Potential operational CONOPs include the (time-varying) roles of different locations, required flying schedules at operating locations, and the types of maneuvers by aircraft. Potential logistical CONOPs, by resource or activity, include the use of a pull versus push system, the extent of mobility for support and storage assets, delivery and transportation capabilities, the potential for lateral resupply, and the echelon and authority of resource providers. Given decisions regarding the CONOPs to be tested, other model inputs can be generally categorized as follows:

- the **resource utilization by location and MDS for each period (*t*)**, which is dependent on the operating location status and mission at that period
- **resource resupply rate and level**, based on the echelon of authority and the provider(s) of resources, as specified in the logistics CONOPs
- **set of operating and support locations**, including the physical location, operational status each period, operational roles and capabilities at each period, resource inventories and storage capacities, throughput capacities, authorization and delivery lead times

- **planning factors**, such as mission characteristics (e.g., ingress and egress time, loiter time), resource characteristics (e.g., size, weight, cost), resource utilization rates based on the MDS and mission characteristics, repair times for maintenance, lead times for spare parts, transportation asset characteristics (e.g., capacity, speed, availability)
- **characteristics of the C2 system**, in particular the extent of C2 disruptions or degradations (quantified as delays in authorization, lead, and delivery times).

An important characteristic of PLATO is its ability to model adversary actions and their impacts on various stages of the logistics OODA loop. Adversary actions may include, but are not limited to, denial of operating and support locations, disruption of C2 information flows and subsequent delays, and loss (or delayed replenishment) of resources and buffer stocks. Note that disruptions due to misinformation, deceptive practices, or loss of interoperability or surveillance and target acquisition capabilities are more difficult to model in this framework. The analyses presented in this report draw on RAND's substantial body of research under the Combat Operations in Denied Environments umbrella,[62] which assesses likely adversary actions and their impacts.

Outputs for Assessing Combat Support to ACE Operations

Having specified the full set of inputs, PLATO evaluates, at each location at each timestep, the resources that are consumed, along with the resources that are available (from both buffer stocks and new deliveries), in order to identify the **resource-limited sortie potential**. This sortie potential is identified and reported for each resource independently; the most limiting resource determines the location's overall potential sorties for that timestep.

As an example, consider Figure 4.2, drawn from a notional example in which 12 sorties from a single MDS are required at a location over a 20-day period. Suppose only three resources are being examined, namely, aircraft availability due to spare parts, fuel, and munitions. In this example, there are sufficient resources to generate the requested sorties on days 1 through 4. Fuel limits the sortie generation capability on days 5, 8, and 11, while both fuel and available aircraft constrain sortie generation on day 6, and both munitions and available aircraft bound sortie generation on day 7. Available aircraft serves as the limiting factor on all other days.

---

[62] Brent Thomas, Mahyar A. Amouzegar, Rachel Costello, Robert A. Guffey, Andrew Karode, Christopher Lynch, Kristin F. Lynch, Ken Munson, Chad J. R. Ohlandt, Daniel M. Romano, Ricardo Sanchez, Robert S. Tripp, and Joseph V. Vesely, *Project AIR FORCE Modeling Capabilities for Support of Combat Operations in Denied Environments*, RAND Corporation, RR-427-AF, 2015.

Figure 4.2. Notional PLATO Outputs

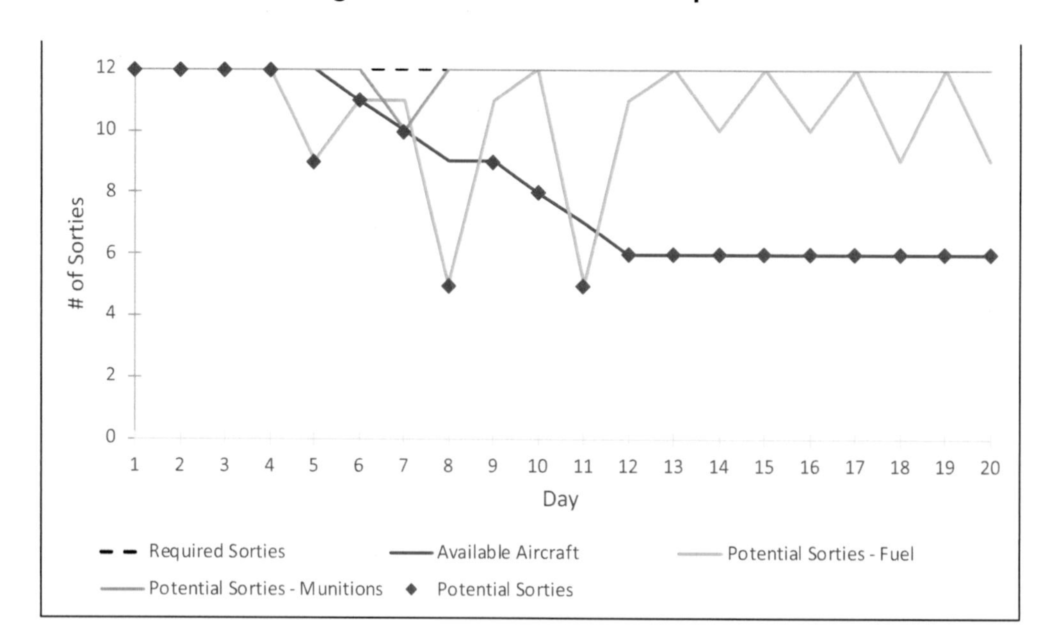

Overview of Mathematical Formulations Used by the PLATO Model

As previously noted, PLATO is implemented using difference equations to capture system dynamics. This approach was used in the model rather than an alternative, such as discrete event simulation, because the dynamics of resource utilization can be faithfully represented by deterministic linear equations and, hence, can be computed more quickly. However, some event-based logic at the end or beginning of each timestep is included in PLATO for execution of aircraft maneuvers, sortie generation, and resupply.

The difference equations on which the model is based take two basic forms, depending on whether the utilized resource represents a stock of reparable inventory, such as spare parts, or inventory that cannot be repaired, such as fuels or munitions. For resource r representing a stock of non-reparable inventory at location i at time $t+1$, the inventory level, denoted $R_i^r(t+1)$, is computed by Equation 4.1 as follows:

$$R_i^r(t+1) = R_i^r(t) - \sum_m \alpha_{im}^r(t) A_{im}^{MC}(t) + \sum_{j \neq i}\left(\beta_{ji}^r\left(t - \delta_{ji}^r\right) - \beta_{ij}^r(t)\right) \qquad (4.1)$$

Here, $\alpha_{im}^r(t)$ is the planned utilization rate for resource r generated by MDS m at location i at time t, $A_{im}^{MC}(t)$ represents the number of mission-capable aircraft for MDS m at location i at time t, $\beta_{ij}^r(t)$ specifies the resupply rate of resource r from location i to location j at time t, and δ_{ji}^r denotes the resupply delay of resource r from location i to location j. Note that this representation corresponds to an unconstrained case, in which the planned sortie generation rate

$\alpha_{im}^{r}(t)$ is assumed to be achieved (we discuss below how resource constraints limit the achievable sorties).

For resource l representing a reparable spare part at location i at time $t+1$, the inventory level, denoted $R_i^l(t+1)$, and the expected part failures, denoted $F_{im}^l(t)$, are respectively computed by Equations 4.2 and 4.3 as follows:

$$R_{im}^l(t+1) = R_{im}^l(t) - F_{im}^l(t) + \sum_{j \neq i} \left(\beta_{jim}^l(t - \delta_{ji}^l) - \beta_{ijm}^l(t) \right) + W_{im}^l(t) \quad (4.2)$$

$$F_{im}^l(t) = q_m^l \lambda_m^l \rho_{im}(t) g_{im}(t) A_{im}^{MC}(t) \quad (4.3)$$

In addition to the parameters described above, these equations also use q_m^l denoting the quantity per aircraft of spare part l on MDS m, λ_m^l representing the expected failure rate of spare part l on MDS m, $\rho_{im}(t)$ specifying the planned sortie rate for MDS m out of location i at time t, $g_{im}(t)$ denoting the average sortie duration for MDS m out of location i at time t, and $W_{im}^l(t)$ representing the part repairs of spare part l on MDS m at location i at time t. Note that the repaired parts are themselves calculated by Equation 4.4, where $1/\mu_m^l$ represents the mean time to repair and $\lceil \cdot \rceil$ denotes the ceiling function:

$$W_{im}^l(t) = F_{im}^l\left(t - \left\lceil 1/\mu_m^l \right\rceil\right) \quad (4.4)$$

Note that the current deterministic version of PLATO tracks *expected* parts inventories. These inventory quantities and resupply quantities are calculated from expected failure rates and allow for fractional part inventories. Moreover, Equation 4.3 implicitly allocates the total flying hours in a timestep over all of the mission-capable aircraft for the corresponding MDS and location.[63]

These resources are then integrated to determine the number of mission-capable aircraft as presented in Equation 4.5, where A_{im}^{Total} denotes the total number of aircraft (both mission-capable and non-mission capable), $\lfloor \cdot \rfloor$ denotes the floor function, and $(\cdot)^+$ denotes the positive-part function. This formulation is consistent with a *full cannibalization* assumption, in which any number of failed parts between zero and q_m^l constitutes a single non-mission capable aircraft.

[63] In the case of sortie rates less than 1.0, this formulation may result in a shortfall of sorties that are generated with respect to the desired number. For example, consider a squadron of 12 aircraft with a required sortie rate of 0.5, which implies that six aircraft are required to fly each timestep. If there are eight aircraft available, then there are sufficient aircraft to meet the required sortie rate and one would expect six sorties to be flown and resource utilization based on these six sorties. However, the current structure of Equation 4.3 does not account for the required number of sorties and instead considers only the sortie rate and available aircraft, resulting in 8*0.5 = 4 sorties generated. Expected resource utilization is subsequently affected. This is a limitation of the model when the specified sortie rate is less than 1.0.

$$A_{im}^{MC}(t+1) = \left\lfloor \min\left\{ A_{im}^{Total}(t), \left(\min_{l} \left(\frac{q_m^l A_{im}^{MC}(t) - \left(F_{im}^l(t) - R_{im}^l(t)\right)}{q_m^l} \right)^+ \right) \right\} \right\rfloor \quad (4.5)$$

The resource-limited sortie generation that can actually be achieved is calculated by solving the following linear program (Equation 4.6), for each period t and location i, where decision variable $P_{im}(t)$ represents the number of sorties that can be achieved for MDS m, and data parameter γ_{im}^r is the **per-sortie** utilization rate for resource r generated by MDS m:

$$\begin{aligned} &\max \sum_m P_{im}(t)\, A_{im}^{MC}(t) \\ &\text{s.t.} \sum_m \gamma_{im}^r(t) P_{im}(t) A_{im}^{MC}(t) \leq R_i^r(t)\ \forall r \\ &\quad P_{im}(t) \leq \rho_{im}(t) \\ &P_{im}(t) \geq 0 \end{aligned} \quad (4.6)$$

Note that to capture the effects of resource constraints on sortie generation, we simply replace $\alpha_{im}^r(t)$ in Equation 4.1 with the resource-constrained value $\gamma_{im}^r(t)P_{im}(t)$ after having solved the linear program at period t, and then the model advances to timestep $t+1$. Furthermore, observe that if inventories are insufficient to meet the planned sortie rate, the model assumes that the sortie requirement will be partially satisfied with the available resources.

Applying the Modeling Framework to the PACAF Challenge

In the companion report, which is not available to the general public, we document the results of applying the PLATO model to the challenges presented by PACAF's ACE CONEMP. Using a set of assumptions related to the operational scenario and planned CS enterprise posture, we examine the performance of the enterprise in a conflict scenario. We then explore different options and mitigation strategies to determine how they might improve enterprise performance. Following the conflict analysis, we apply the framework to a notional ACE maneuver scheme that might be used to demonstrate USAF's ability to execute the ACE concept as a means of avoiding conflict or to provide insights on the ability of PACAF to engage in a "fight tonight" scenario.

In this same report, we explore the performance of the CS enterprise under two ACE operational cases: one based on an expected operational scenario in conflict and the other on a competition state exercise ACE maneuver scheme. We use an integrated modeling framework to assess the implications of adjustments to the assumptions of these analyses and the overall performance of the enterprise designed for efficiency. Additionally, we explore several different mitigation strategies to enhance enterprise performance to demonstrate both the flexibility of the modeling framework and the potential value of these strategies.

Chapter 5. Findings and Recommendations

USAF's CS enterprise is a complex system with diverse stakeholders and resources. Designed for efficiency, it relies on timely information about resource status and requirements, assured transportation, appropriately positioned inventories, sophisticated forecasting methods, accurate forecasting planning factors, and assessments of planned operations. Personnel responsible for sustaining operations at forward operating locations and within the theater are often at the mercy of enterprise-level decisionmakers setting resource levels in competition and making allocation and reallocation decisions in conflict. Operating in competition absent the fog of war, the CS enterprise, for various reasons, cannot meet the 80-percent readiness rate directed by the Secretary of Defense in 2018.[64] The current state does not bode well for the prospect in conflict where state-of-the-world uncertainties are likely.

In conflict, adversary actions can render many parts of the enterprise designed around efficiency unstable. Degradation of communication nodes can reduce knowledge of available inventories or resource needs. Attacks on lines of communication can disrupt assured transportation. Operational demands could exceed what was planned. There are strategies that USAF can pursue to mitigate these uncertainties at the enterprise level, as well as actions that PACAF can take relative to its theater of operations and those enterprise elements over which they have control and influence.

Key Findings

Our analysis revealed the following:

- The CS enterprise, built on efficiency, relies on the timely communication of asset requirements and assured and responsive transportation, which today fails to achieve desired readiness rates and will likely be more challenged in a conflict with a near-peer adversary.[65]
- Processes for communicating resource status and replenishment needs will aid in mitigating CS enterprise disruptions resulting from adversary attacks; however, other actions will likely be necessary to support operational missions in a contested environment.

[64] See Aaron Mehta, "Mattis Orders Fighter Jet Readiness to Jump to 80 Percent—in One Year," *Defense News*, October 9, 2018. We note that resource availability is just one part of readiness. Manpower, training, and repair activities also factor into readiness rates.

[65] This observation corresponds to an observation by Snyder et al. (2021), but it is further amplified by considering it in the context of the logistics OODA loop.

- Operating the CS enterprise in conflict will require logisticians above the unit level to be fully aware of planned sortie demand and to have the ability to forecast required replenishment based on that demand.
- Planning factors used to make posture decisions during competition may not accurately reflect the expected intensity of operations in a conflict with a near-peer adversary.

Recommendations

In light of our findings from this analysis, we recommend that USAF consider the following mitigation strategies:

- Each PACAF CS functional area, in coordination with its enterprise functional community, should develop and practice TTP for executing the logistics support plan in a communications-degraded environment.
- PACAF should adopt a methodology to determine which assets in the CS enterprise should be pushed to forward operating locations and the rate and quantities that should be pushed if it becomes necessary to temporarily shift from a pull to a push system in a communications-degraded environment. A proposed basis for such a methodology is provided in Chapter 4 of this report.
- USAF and PACAF should consider a "partial buyout" buffer stock strategy to mitigate expected resource shortages for planned operations.
- USAF should review the WMP-5 planning factors used to compute RSPs to ensure these factors reflect the intensity of operations outlined in PACAF's OPLAN, including expected attrition.

Appendix. Combat Support Enterprise Decision Authorities

Combat operators are defining methods and authorities for continuing operations within a contested and communications-degraded environment (e.g., mission type orders, baseline ATOs, and guidance packages). Likewise, CS communities are beginning to think through how processes could continue in a high-end fight because CS operations have no standard way of communicating authorities if the typical processes become infeasible. During competition, decision authorities for CS at the enterprise and theater levels are well defined.[66] Published directives, including DoD Manuals and AFIs, outline enterprise-wide authorities and responsibilities for USAF resource and personnel management. In this appendix, we review these authorities and evaluate how decisionmaking may adapt during conflict when C2 could shift from the more centralized enterprise level to more-dispersed levels, such as theater, base, or unit.

Decision Authorities During Competition

In this analysis, we focus on five support functions—munitions, fuels, spare parts, repair capability, and transportation—that are essential to mission generation capability and the execution of OPLANs. During competition, these functions have well-defined, routinely practiced processes, as outlined in the next sections. We do not present a comprehensive review of each functional area but rather focus on how requirements are determined, which organizations have decision authorities, and the scope of those authorities.[67]

We focus on the decision authorities during competition for two reasons. First, decisions made by stakeholders in the CS enterprise during competition directly affect enterprise performance in conflict. If resources are not positioned in the right locations and in the right quantities during competition, the enterprise runs the risk of failure in conflict. Second, the process of practicing and exercising ACE maneuver operations during competition can serve as a deterrent for conflict. If resource allocation decisions made in support of ACE exercises fail to sufficiently support the demonstration of USAF ACE capabilities, deterrence is at risk.

---

[66] Joint Doctrine Note 1-19, *Competition Continuum*, Joint Chiefs of Staff, June 3, 2019.

[67] For more-detailed descriptions of CS processes and other functional areas, see James Leftwich, Robert Tripp, Amanda Geller, Patrick Mills, Tom LaTourrette, C. Robert Roll, Jr., Cauley Von Hoffman, and David Johansen, *Supporting Expeditionary Aerospace Forces: An Operational Architecture for Combat Support Execution Planning and Control*, RAND Corporation, MR- 1536-AF, 2002; Patrick Mills, Ken Evers, Donna Kinlin, and Robert S. Tripp, *Supporting Air and Space Expeditionary Forces: Expanded Operational Architecture for Combat Support Execution Planning and Control*, RAND Corporation, MG- 316-AF, 2006; and Kristin F. Lynch, John G. Drew, Robert S. Tripp, Daniel M. Romano, Jin Woo Yi, and Amy L. Maletic, *An Operational Architecture for Improving Air Force Command and Control Through Enhanced Agile Combat Support Planning, Execution, Monitoring, and Control Processes*, RAND Corporation, RR-261-AF, 2014.

Munitions

Munitions are managed globally using well-defined forecasting, authorization, and allocation processes initiated and led by the global ammunition control point (GACP) at Hill Air Force Base, Utah. Forecasting begins with theater munitions planners using OPLANs to determine which targets need to be prosecuted to achieve the CCDR's intent and which type of munition is best suited for each target set. CCMDs submit annual theater-level requirements forecasts to the GACP as inputs into the Non-nuclear Consumable Annual Analysis process. The Air Force Deputy Chief of Staff, Plans and Programs, Operational Capability Requirements Directorate (AF/A5R) validates theater requirements for each CCMD and determines appropriate primary, secondary, and tertiary munitions for acquisition. This is the unconstrained global munitions requirement.[68]

DoD, however, does not resource to the full, unconstrained munitions requirement. To set munitions authorizations and priorities, the GACP leads two working groups, the Budget Buy working group and the Munitions Allocation working group.[69] These working groups use the theater-level forecasts to prioritize the acquisition, allocation, and positioning of the munitions globally.

Munitions are allocated to each theater (with some remaining in a worldwide stockpile to be allocated and repositioned as needed to meet emerging or enduring requirements). The theater-allocated munitions are managed by the C-MAJCOM staff within that theater and positioned across the theater to meet OPLAN requirements and the CCDR's intent within net explosive weight restrictions.[70]

Fuels

Like munitions, fuel resources are managed globally. DLA Energy is responsible for fuel procurement and contracting, and the Air Force Petroleum Agency is responsible for centralized support functions. During competition, fuel demand is largely determined by the number of hours each type of aircraft is expected to fly that year, as documented in the flying hour program. USAF provides fuel estimates, by theater, through the CCMD's JPO, based on flying hour

---

[68] The formal munitions requirements process is outlined in the Peacetime Conventional Ammunitions Requirements memorandum, which is coordinated and published by the GACP. See Department of Air Force Manual (DAFMAN) 21-201, *Munitions Management*, Department of the Air Force, May 3, 2022; and Air Force Manual (AFMAN) 11-212, *Requirements for Aircrew Munitions*, Department of the Air Force, June 25, 2020.

[69] DAFMAN 21-201, 2022, pp. 260-262.

[70] The C-MAJCOM staff is responsible for the organize, train, and equip (OT&E) service component activities for that MAJCOM. The AFFOR staff has the responsibility during operations in support of the CCMD. However, initiatives to reduce the size of headquarter-level staff and other manpower reductions have necessitated many personnel to wear two hats as both MAJCOM staff and AFFOR staff, responsible for both OT&E and warfighting.

program requirements.[71] However, there is a congressional limit on the total amount of fuel that can be on hand, globally, at any given time. The Assistant Secretary of Defense for Sustainment is ultimately responsible for managing policies to comply with the congressional mandate; DLA executes in accordance with those policies.[72] In-theater inventories are determined by stockage objectives set by the CCMD JPO as the necessary inventory for meeting mission requirements and fuel is stored across the theater to meet flying hour program requirements.[73] Each stage of the fuel requirements, authorization, and priority-setting process is worked with and determined by the CCMD JPO in coordination with base-level fuel functional personnel and C-MAJCOM staff.

Spare Parts

Similar to fuels, spare parts requirements are partially determined by the expected annual flying hour program in conjunction with historical consumption data. Air Force Materiel Command (AFMC) logistics systems incorporate annual estimated flying hours and historical consumption data to forecast part failure rates, which are used to determine a worldwide requirement, similar to munitions.[74] USAF uses a readiness-based level (RBL) process to allocate the worldwide requirement across bases and depots, with the goal of minimizing expected backorders. Entered into the enterprise spares management system (Integrated Logistics System-Supply [ILS-S]), the RBL is the base stockage level.

However, unlike fuels and munitions, there are several global managers of spare parts. Some parts are organically procured (by USAF item managers), repaired (by USAF depots), and managed with USAF using the process discussed above. Others, such as consumables, parts common across services, and some others are procured and managed by DLA. And finally, for weapon systems and equipment supported through contractor logistics support (CLS) contracts, the procurement, repair, and management of CLS parts are typically handled by that contractor (e.g., Lockheed Martin and Boeing for the F-22 air vehicle).

For items managed by DLA, USAF uses the Customer Oriented Leveling Technique (COLT) model to set base stockage levels, with the goal of minimizing customer wait time. COLT levels are entered into ILS-S as the RBL for DLA-managed parts.

Spare parts for CLS-supported aircraft are part of the CLS contract and are typically negotiated into the contract as a level of availability rather than a stockage policy. For example, USAF could negotiate a contract for 90-percent aircraft availability, meaning the aircraft could

---

[71] AFI 11-102, *Flying Hour Program Management*, Department of the Air Force, December 8, 2020, p. 3.

[72] Department of Defense Manual (DoDM) 4140.25, Volume 5, *DoD Management of Energy Commodities: Support of Joint Operations, Contingencies, Civil Authorities, and Military Exercises*, U.S. Department of Defense, March 2, 2018, incorporating change 2, April 4, 2019, p. 4.

[73] AFI 23-201, *Fuels Management*, Department of the Air Force, August 9, 2021, p. 42.

[74] Department of Air Force Instruction 23-101, *Materiel Management,* July 8, 2021, p. 73.

be down (not available) for parts or maintenance up to 10 percent of the time. It is the contractor's responsibility to calculate, procure, and maintain the parts needed to ensure the aircraft are available 90 percent of the time. With this sort of agreement, USAF has limited input and visibility into the contractor stockage policies and must rely on the contractor to meet the purchased level of aircraft availability.

In addition to base-level supply, each weapon system is provisioned an RSP to use when it is deployed. RSPs are designed and assembled to provide spare parts support for a limited time horizon based on wartime missions outlined in the WMP-5.[75] The lead MAJCOM and AFMC are responsible for the total force RSP requirement, and they determine spare part priorities. The Director of Logistics, Civil Engineering, Force Protection and Nuclear Integration (AFMC/A4), with DLA, is responsible for the long-term management and supply of spare parts for RSPs.

Repair Activities

USAF sorts aircraft repair activities, or maintenance, into three levels: organizational (at the base, on the flightline); intermediate (backshops on base or centralized in a region); and depot. The decision authorities for repair activities vary by level of maintenance and whether the aircraft is maintained organically or through CLS.[76]

Lead MAJCOMs, through Headquarters USAF, Director of Current Operations (HAF/A3O), use the flying hour program to determine *wing-level maintenance requirements*, that is, the expected amount of on-base, flightline, and backshop repair activities for each weapon system on a base. Wing-level maintenance manpower requirements are calculated using the Logistics Composite Model (LCOM) and workforce standards based on wing-level repair requirements.[77] Total wing-level maintenance authorizations are negotiated by career field managers and the lead unit. The base-level unit reviews and coordinates the unit manpower document that determines the authorized workforce for wing-level repair.[78] Repair workloads are forecast across the fleet,

---

[75] Department of Air Force Instruction 23-101, 2021, p. 104.

[76] CLS weapon systems (such as the F-35) are supported through similar maintenance activities with maintenance data from the aircraft sent directly to the original equipment manufacturer (OEM) through a portable maintenance aid (PMA). Once parts are received, the OEM provides repair instructions (through the PMA) to the base-level maintainers. As needed, field support teams from the OEM are sent to the base to assist with maintenance activities.

[77] LCOM was developed by RAND and the Air Force Logistics Command in the late 1960s. It simulates base support, including aircraft maintenance, to determine workforce requirements for maintenance units (R. R. Fischer, W. F. Drake, J. J. Delfausse, A. J. Clark, and A. L. Buchanan, *The Logistics Composite Model: An Overall View*, RAND Corporation, RM-5544-PR, 1968; Edward Boyle, *LCOM Explained*, Logistics and Human Factors Division, Air Force Human Resources Laboratory, July 1990).

[78] AFI 21-101, *Aircraft and Equipment Maintenance Management*, Department of the Air Force, January 16, 2020, p. 33.

but ultimately the wing commander determines base-level repair priorities according to maintenance repair workload, on-hand parts, and the available maintenance manpower.[79]

Depot-level workloads are also determined by the flying hour program and any additional workload authorized by the system program office, such as weapon system modifications. Similar to wing-level maintenance requirements, depot-level workforce requirements are determined using manpower standards. AFMC and the Air Logistics Complexes use the Execution and Prioritization of Repair Support System (EXPRESS) to determine depot-level workload priorities.[80]

Transportation

Transportation decisionmaking authorities are divided into what is managed by the U.S. Transportation Command (USTRANSCOM) at the enterprise level and what is managed by the CCMD at the theater level. At the enterprise level, strategic transportation assets (e.g., airlift, refuelers, sealift, and ground transportation) are used to meet validated deployment and distribution requirements. USTRANSCOM is responsible for planning (the timeline and the route) and allocating (the mode and the asset) of transportation resources in accordance with the requirements and priorities determined by the priority movement system established by the CJCS. The joint movement board (chaired by the CJCS but rarely used) can arbitrate across competing demands and priorities as needed.

Theater-level movement requirements are determined and prioritized according to the CCDR's intent. The theater deployment and distribution operations center (DDOC) coordinates and synchronizes all modes of movement within the theater. The air mobility division within the AOC plans and executes missions for theater-assigned air assets, coordinating with the tanker and airlift control center and the DDOC.

Decision Authorities During Conflict

When operations shift from competition to conflict, many of the CS functional processes do not change. What does change is the requirement for support (i.e., the resource demand and location based on conflict operations), as well as the associated decision authorities and scope of these authorities as they are shifted to more-forward entities. The CCDR has authority over

---

[79] AFI 21-101, 2020, pp. 21–22.

[80] EXPRESS is a tool that uses daily data, such as available parts and available manpower, to determine workload prioritization to meet availability goals for each weapon system. EXPRESS uses prioritization algorithms developed by RAND in the Distribution and Repair in Variable Environments tool, which determined repair prioritization by maximizing the probability of meeting availability objectives. See David R. Williams, *Examining EXPRESS with Simulation*, thesis, Air Force Institute of Technology, Department of the Air Force Air University, 2012, p. 8; and John B. Abell, Louis W. Miller, Curtis E. Neumann, and Judith E. Payne, *DRIVE (Distribution and Repair in Variable Environments): Enhancing the Responsiveness of Depot Repair*, RAND Corporation, R-3888-AF, 1992.

assets assigned to their theater. The CCDR coordinates through the CCMD staff with the commander of Air Force forces (COMAFFOR) and the AFFOR staff for USAF-specific resources and capabilities. USAF units relay critical information to the AFFOR staff, such as expenditures, usage, and break rates. The AFFOR staff coordinates with the AOC to provide up-to-date information about available assets and resources. The AOC publishes a daily ATO that contains the missions for that day (as well as plans for the following two days).[81] The CCMD, in coordination with the COMAFFOR and the AFFOR staff, can allocate and reallocate theater-assigned resources as needed to meet the published ATO. When in-theater resources are insufficient, the CCDR can request more resources from global assets stockpiles or from other theaters' resources. In the sections below we outline, by functional area, any differences in how requirements are determined, which organizations have authorities, and the scope of those authorities during conflict as compared with operations during competition.

Munitions

Munitions are prepositioned in each theater during peacetime in a clearly documented process, outlined in the previous section on this topic. The CCDRs have authority over munitions allocated to their theaters both during competition and in conflict. Within the theater, the C-MAJCOM and AFFOR staff use CCMD guidance to set munition priorities and allocate munitions within net explosive weight restrictions. If munitions need to be moved intra-theater or intra-command to support a conflict, the AFFOR theater/regional ammunition control point will direct the GACP to initiate a redistribution order in the Theater Integrated Combat Munitions System.[82] If theater needs during conflict exceed the theater allocation, the AFFOR staff will submit a request to the GACP for additional supply from the global inventory.

Fuels

The fuels appendix in the base support plan (BSP) or expeditionary site plan (ESP) contains the estimated contingency fuel requirements for the base.[83] The BSP and ESP, developed by fuel management teams, use time-phased force and deployment data and the WMP-5 to inform the planned fuel consumption during a conflict.[84] DLA provides centralized contingency fuel planning and support to components and monitors bulk petroleum. As in competition, the CCMD JPO, in coordination with base-level fuels support personnel and the AFFOR staff, is responsible for fuel requirements, authorizations, and determining priorities in the theater during a conflict. If

---

[81] For example, for an offensive mission, the ATO contains the aircraft assigned to the mission, targets to be serviced as part of the mission, and munitions to be expended against those targets.

[82] DAFMAN 21-201, 2022, p. 229.

[83] AFI 23-201, 2021, p. 76.

[84] AFI 23-201, 2021, p. 44.

necessary, DLA Energy can authorize local fuel purchases.[85] When demand for fuel resources exceeds the allocation to a theater, the joint materiel priorities and allocation board, acting on behalf of the CJCS, adjudicates and sets the priorities for allocation across the global enterprise.[86]

Spare Parts

Units use their RSPs when deployed in a conflict. Spare parts resupply is estimated according to the flying schedule published in the ATO. However, the actual need for parts is determined when aircraft break, which can corollate with ATO expectations or, in some instances, can be very different from those expectations. AFMC/A4, with DLA and the OEM for CLS support, are responsible for the long-term supply and management of spare parts. Within the theater, the AFFOR staff, with CCMD guidance, sets priorities for spare parts allocations. Lead commands or the MAJCOM can request a contingency high priority mission support kit (CHPMSK) through AFMC. A CHPMSK provides additional resources to augment the RSP and reallocates base supply from other locations to support contingency operations.[87]

Repair Activities

As during competition, during conflict at the organizational level, repair requirements are affected by flying hours (determined by the ATO) and the frequency of part failures. The wing uses CCMD guidance to set repair priorities at the organizational level.

However, repair requirements during conflict are also directly affected by adversary actions, that is, the damage adversary actions inflict on aircraft. The Aircraft Battle Damage Repair (ABDR) program is globally managed by AFMC and is typically only used during contingency operations when the speed of repair is critical.[88] Other depot-level repair authorizations and priorities are determined by AFMC and the CCMD.

Transportation

Transportation and distribution of resources in theater during conflict are determined by needs identified in the CCMD's plan. Similar to transportation authorities in competition, USTRANSCOM is responsible for planning and allocating transportation resources for inter-theater transportation and for intra-theater transportation provided by assets not assigned to the theater. The movement of theater-assigned assets is determined and prioritized according to the CCDR's intent with coordination through the DDOC. During conflict, assets may be *chopped*

---

[85] DoDM 4140.25, Volume 5, 2019, p. 6

[86] DoDM 4140.25, Volume 5, 2019, p. 7.

[87] AFI 23-101, 2020, p. 17.

[88] The ABDR program focuses less on longer-term durability issues, such as corrosion.

(i.e., change of operational control assets) from their typical points of authority to a theater for theater use. Chopped assets are treated as theater-assigned assets; their movement and prioritization are authorized by the CCMD.

During conflict, the CCMD may establish standard theater airlift routes (STARs) using theater-assigned assets. Establishing STARs early in a conflict can reduce cargo backlogs at transshipment points and establishes a resupply schedule for forward units. As with other theater-assigned assets, aircraft flying STARs are prioritized according to the CCDR's intent.

Decision Authorities During Contested, ACE-Like Operations in Conflict

In a near-peer conflict where the adversary is capable of a multi-domain attack, USAF operational planners look to maneuver as a way to protect forces and complicate adversary targeting. By *maneuver* we are referring to the movement of aircraft and the supporting ground forces either proactively (before an attack) or reactively (during or after an attack).[89] A multi-domain attack by the adversary could include both kinetic and non-kinetic offensive operations resulting in degraded or denied communications in the area of responsibility. If communications are degraded or denied, a unit may not receive a new ATO every day and forward operating units may have to develop mission plans using baseline ATOs and guidance packages of their own. In this section, we analyze functional responsibilities and authorities when forces are operating in a contested, ACE-like environment—maneuvering, potentially without communications capabilities.

Munitions

During contested operations, if communications are degraded or denied and a unit does not receive a daily ATO, forward munitions personnel may have to work with forward operational units to develop mission plans that support the CCDR's intent using available resources. If unit munitions personnel are unable to update the AFFOR staff on the number of munitions expended and the on-hand inventory, the AFFOR staff may have to predict munitions resupply requirements for the forward unit. Because the AFFOR staff would have the last inventory status report (before communications were lost) and know the aircraft type, standard configuration load, and expected expenditure rates according to the last ATO, the AFFOR staff could forecast expected usage and estimate the current munitions inventory. The AFFOR staff could also approximate the loss from attrition, that is, munitions destroyed by adversary kinetic attack based on the situational reports from the forward location.

By evaluating expected inventory levels for forward locations when communications are denied, the AFFOR staff would be prepared to push munitions to those locations in the event that

[89] See PACAF, 2020.

communications were not restored before munitions inventories were expected to reach critical levels. The AFFOR staff could then work with the CCMD staff to allocate or reallocate theater-owned munitions to the forward location based on projected usage. Or, if the CCMD does not have enough theater-assigned assets, it could request additional munitions resupply from the global inventory.

Key to maintaining resources at the levels needed to conduct operations is the AFFOR staff's visibility into on-hand inventory and expected expenditures. Establishing a battle rhythm for when the AFFOR staff will receive these updates will simplify deciding when the AFFOR staff needs to switch from monitoring status reports to calculating needed resupply.

Fuels

Similar to munitions, if communications are degraded and a unit does not have a daily ATO, forward fuels personnel would need to work with forward operational units to schedule missions that meet the CCDR's intent, constrained by the available fuel resources. If unit fuels personnel are unable to update the JPO on the amount of fuel consumed and the on-hand inventory, the JPO would have to rely on previous fuel consumption forecasts or conduct consumption calculations independently. The JPO could use such information as the aircraft type, sortie rate, and sortie duration from the last ATO to calculate expected fuel consumption. Drawing on other reports received by the AFFOR staff about the forward location, the JPO could estimate fuel lost to attrition. Then, knowing the reserve fuel factor for the location, the JPO could calculate expected remaining fuel resources to better understand when that location would need resupply.

To maintain adequate fuel resources to conduct missions, the JPO needs to have visibility into on-hand inventory and expected consumption. Like munitions, by evaluating and estimating remaining fuels as soon as communications are lost, the JPO would be prepared to push additional fuel (or reduce fuel deliveries if a forward base is damaged or no longer being used) before inventories were expected to reach critical levels.

Spare Parts

Like munitions and fuels, the need for spare parts is based on consumption, but unlike other resources, spare parts consumption is not easily predicted because aircraft break at different times for different reasons. The RSP is a forward-deployed unit's source for spare parts, assuming no connectivity to the CS network for the initial phase of operations. Once those stocks are depleted (or, in the case of parts that are not included in an RSP, once the first part breaks), the unit needs to be resupplied.

Aircraft damaged during conflict add another layer of complexity to spare parts challenges. An aircraft damaged by adversary attack may need parts for repair that are not typically in an RSP, or it may need parts that are typically so reliable that those parts have a very low historical break rate. For parts with a low break rate, it may be difficult to forecast a resupply rate. On the other hand, battle-damaged aircraft that are attritted by adversary attack could serve as a source of supply for other undamaged aircraft. Maintainers could cannibalize parts off the battle-damaged aircraft and use them as spares to maintain other serviceable aircraft.

With limited or no communications, the unpredictability of breaks makes anticipating what spare parts a forward unit may need during conflict challenging. There are, however, models that can help forecast the spare parts needed. With information from the ATO, such as MDS, sortie rate, and sortie duration, the AFFOR staff or analysts at the 635th Supply Chain Operations Wing could calculate anticipated demands using existing systems, such as the Aircraft Availability Model or the Aircraft Sustainability Model.[90] The resupply of parts could come from other forward units, base supply in the rear, or the enterprise supply chain.

Repair Activities

Like the resupply of spare parts, the repair activity needed during conflict is unpredictable as aircraft break at different times for different reasons. Past experience has shown, however, that aircraft flying in combat tend not to break as frequently as during noncombat operations. This may be because the aircraft are flying more each day, or it may be because pilots are more willing to ignore noncritical breaks, such as a radio channel that has some static. During competition, radio static would require repair. During combat, pilots may choose to use a different channel and ignore the repair until the aircraft has a mission-critical break requiring repair.

The unpredictability of aircraft breaks makes anticipating repair workload at a forward operating location challenging. During competition, the wing commander determines repair priorities given the current status of parts, personnel, and needed repairs. In conflict, units may not be collocated with their wing commander, so the location commander, whether that is a base operating support integrator, a lead squadron commander, or someone else, should be making repair prioritization decisions. As in competition, those decisions are going to depend on maintenance repair workload, on-hand parts, and the available maintenance personnel.

Battle-damaged aircraft, damaged through adversary attack, may require more repair expertise than unit-level maintainers have at the forward location. Thus, the need for depot-level repair capabilities may be large in a near-peer conflict. Although the exact damage cannot be forecast, depot-level repair teams should be established and positioned to provide support when needed. It would be beneficial to position depot repair teams at each operating location, but that would be cost-prohibitive. As a result, depot repair teams will have to be shared across locations. Although the depot teams are from AFMC, the CCDR (through the CCMD staff and the AFFOR staff) should be setting depot-level repair priorities for repair activities within the theater.

---

[90] The Aircraft Availability Model is a USAF tool used to calculate resupply needs of spare parts during peacetime. The Aircraft Sustainability Model is a USAF tool used to calculate spare parts requirements.

Transportation

Strategic airlift aircraft and refuelers are considered high-value assets by USAF. It is likely that, in a contested environment, these assets will limit operations to areas that are relatively safe from adversary attack. If these strategic assets operate mainly in the rear, it is unlikely they will lose communications capabilities and operations will continue as outlined in the earlier section discussing transportation during conflict.

For theater-assigned assets, they will likely have to operate closer to the front lines in order to resupply forward locations. However, they too are considered high-value assets (there are not many of them assigned to each theater) and are unlikely to lose communications capabilities. If communications capabilities are lost between the main operating base and forward locations, it may be these theater-assigned airlift assets that provide the missing communications link. They could transmit a new ATO to the forward operating location (e.g., deliver a thumb drive or disk with the new ATO on it). In return, they can transmit location and resource status back to the main operating location, such as the number of aircraft damaged, amount of fuel available, and status of the runway.

Establishing STARs early in a conflict would allow forward units to know when resupply or information (if communications were denied or degraded) would be available to that unit. On the other hand, establishing a set STAR schedule would present the adversary with a predictable target.

Abbreviations

ABDR	Aircraft Battle Damage Repair
ACE	agile combat employment
AFFOR	Air Force forces
AFI	Air Force Instruction
AFMC	Air Force Materiel Command
AFMC/A4	Director of Logistics, Civil Engineering, Force Protection and Nuclear Integration, Air Force Materiel Command
AOC	air operations center
ASD	average sortie duration
ATO	air tasking order
C2	command and control
CCDR	combatant commander
CCMD	combatant command
CJCS	Chairman of the Joint Chiefs of Staff
CLOUT	Coupling Logistics to Operations to meet Uncertainty and the Threat
CLS	contractor logistics support
C-MAJCOM	component major command
CONEMP	concept of employment
CONOP	concept of operations
CONUS	continental United States
CS	combat support
DAF	Department of the Air Force
DDOC	deployment and distribution operations center
DLA	Defense Logistics Agency
DoD	U.S. Department of Defense
DoDM	Department of Defense Manual
EXPRESS	Execution and Prioritization of Repair Support System
FED LOG	Federal Logistics Data
GACP	global ammunition control point
ILS-S	Integrated Logistics System-Supply
JADC2	joint all-domain command and control
JLEnt	joint logistics enterprise
JPO	joint petroleum office
LCOM	Logistics Composite Model

MAJCOM	major command
MDS	mission design series
MICAP	mission impaired capability awaiting parts
NSN	national stock number
OEM	original equipment manufacturer
OODA	observe-orient-decide-act
OPLAN	operation plan
OPTEMPO	operations tempo
PACAF	Pacific Air Forces
PACE	primary, alternate, contingency, and emergency
PAF	RAND Project AIR FORCE
PLATO	Persistent Logistics Analysis Tool and Optimization
RBL	readiness-based level
RSP	readiness spares package
STAR	standard theater airlift route
TTP	tactics, techniques, and procedures
USAF	U.S. Air Force
USTRANSCOM	U.S. Transportation Command
WMP-5	War Mobilization Plan–Volume 5
WRM	war reserve materiel

References

Abell, John B., Louis W. Miller, Curtis E. Neumann, and Judith E. Payne, *DRIVE (Distribution and Repair in Variable Environments): Enhancing the Responsiveness of Depot Repair*, RAND Corporation, R-3888-AF, 1992. As of September 3, 2021: https://www.rand.org/pubs/reports/R3888.html

AFI—*See* Air Force Instruction.

AFMAN—*See* Air Force Manual.

Air Force Doctrine Publication 4-0, *Combat Support*, LeMay Center for Doctrine, January 5, 2020.

Air Force Instruction 11-102, *Flying Hour Program Management*, Department of the Air Force, December 8, 2020.

Air Force Instruction 21-101, *Aircraft and Equipment Maintenance Management*, Department of the Air Force, January 16, 2020.

Air Force Instruction 23-201, *Fuels Management*, Department of the Air Force, August 9, 2021.

Air Force Manual 11-212, *Requirements for Aircrew Munitions*, Department of the Air Force, June 25, 2020.

Bhattacharya, D. K., "On Multi-Item Inventory," *European Journal of Operational Research*, Vol. 162, No. 3, May 2005.

Boyle, Edward, *LCOM Explained*, Logistics and Human Factors Division, Air Force Human Resources Laboratory, July 1990.

Chairman of the Joint Chiefs of Staff Instruction 4310.01F, *Logistics Planning Guidance for Pre-Positioned War Reserve Materiel*, Joint Chiefs of Staff, August 29, 2022.

Cohen, I. K., John B. Abell, and Thomas F. Lippiatt, *Coupling Logistics to Operations to Meet Uncertainty and the Threat (CLOUT): An Overview*, RAND Corporation, R-3979-AF, 1991. As of September 3, 2021: https://www.rand.org/pubs/reports/R3979.html

Coram, Robert, *Boyd: The Fighter Pilot Who Changed the Art of War*, Back Bay Books/Little, Brown and Company, Hachette Book Group, 2002.

DAF—*See* Department of the Air Force.

DAFMAN—See Department of Air Force Manual.

Defense Logistics Agency, "FED LOG–Federal Logistics Data," database, undated. As of July 1, 2021:
https://www.dla.mil/Information-Operations/Services/Applications/FED-LOG/

Defense Logistics Agency, Analytics Center of Excellence, "DLA Distribution Receipts," monthly data feed to RAND, January 2021.

Department of Air Force Instruction 23-101, *Materiel Management,* Department of the Air Force, July 8, 2021.

Department of Air Force Manual 21-201, *Munitions Management*, Department of the Air Force, May 3, 2022.

Department of Defense Manual 4140.25, Volume 5, *DoD Management of Energy Commodities: Support of Joint Operations, Contingencies, Civil Authorities, and Military Exercises*, U.S. Department of Defense, March 2, 2018, incorporating change 2, April 4, 2019.

Department of the Air Force, *War and Mobilization Plan 2011*, Vol. 5: *Basic Planning Factors and Data*, October 2010.

DLA—*See* Defense Logistics Agency.

DoD—*See* U.S. Department of Defense.

DoDM—*See* Department of Defense Manual.

Dunford, Joseph, F., Jr., "Institute for Defense Analysis Study," information memorandum to the Secretary of Defense, April 30, 2018.

Fischer, R. R., W. F. Drake, J. J. Delfausse, A. J. Clark, and A. L. Buchanan, *The Logistics Composite Model: An Overall View*, RAND Corporation, RM-5544-PR, 1968. As of September 3, 2021:
https://www.rand.org/pubs/research_memoranda/RM5544.html

Hastings, Katherine C., James A. Leftwich, Vikram Kilambi, and Ronald G. McGarvey, *Maneuvering Beyond Support: Application of the PLATO Model to Determine Sufficiency of Munitions, Fuel, and Spare Parts Required to Support a Pacific Maneuver Operation*, RAND Corporation, forthcoming, Not available to the general public.

Joint Doctrine Note 1-19, *Competition Continuum*, Joint Chiefs of Staff, June 3, 2019.

Joint Publication 4-0, *Joint Logistics*, Joint Chiefs of Staff, February 4, 2019, incorporating change 1, May 8, 2019.

Karimi, Jahangir, and Benn R. Konsynski, "Globalization and Information Management Strategies," *Journal of Management Information Systems*, Vol. 7, No. 4, Spring 1991.

Leftwich, James, Robert Tripp, Amanda Geller, Patrick Mills, Tom LaTourrette, C. Robert Roll, Jr., Cauley Von Hoffman, and David Johansen, *Supporting Expeditionary Aerospace Forces: An Operational Architecture for Combat Support Execution Planning and Control*, RAND Corporation, MR-1536-AF, 2002. As of August 31, 2021: http://www.rand.org/pubs/monograph_reports/MR1536.html

Lynch, Kristin F., Anthony DeCicco, Bart E. Bennett, John G. Drew, Amanda Kadlec, Vikram Kilambi, Kurt Klein, James A. Leftwich, Miriam E. Marlier, Ronald G. McGarvey, Patrick Mills, Theo Milonopoulos, Robert S. Tripp, and Anna Jean Wirth, *Analysis of Global Management of Air Force War Reserve Materiel to Support Operations in Contested and Degraded Environments*, RAND Corporation, RR-3081-AF, 2021. As of February 15, 2023: https://www.rand.org/pubs/research_reports/RR3081.html

Lynch, Kristin F., John G. Drew, Robert S. Tripp, Daniel M. Romano, Jin Woo Yi, and Amy L. Maletic, *An Operational Architecture for Improving Air Force Command and Control Through Enhanced Agile Combat Support Planning, Execution, Monitoring, and Control Processes*, RAND Corporation, RR-261-AF, 2014. As of August 31, 2021: http://www.rand.org/pubs/research_reports/RR261.html

Mehta, Aaron, "Mattis Orders Fighter Jet Readiness to Jump to 80 Percent—in One Year," *DefenseNews*, October 9, 2018.

Mills, Patrick, Ken Evers, Donna Kinlin, and Robert S. Tripp, *Supporting Air and Space Expeditionary Forces: Expanded Operational Architecture for Combat Support Execution Planning and Control*, RAND Corporation, MG- 316-AF, 2006. As of August 31, 2021: http://www.rand.org/pubs/monographs/MG316.html

Novotná, Veronika, "Numerical Solution of the Inventory Balance Delay Differential Equation," *International Journal of Engineering Business Management*, Vol. 7, 2015.

Office of the Chairman of the Joint Chiefs of Staff, *DOD Dictionary of Military and Associated Terms*, Joint Staff, November 2021.

PACAF—*See* Pacific Air Forces.

Pacific Air Forces, "Agile Combat Employment (ACE): PACAF Annex to Department of the Air Force Adaptive Operations in Contested Environments," Department of the Air Force, June 2020.

Simchi-Levi, David, *Operations Rules: Delivering Customer Value Through Flexible Operations*, MIT Press, 2013.

Snyder, Don, Kristin F. Lynch, Colby Peyton Steiner, John G. Drew, Myron Hura, Miriam E. Marlier, and Theo Milonopoulos, *Command and Control of U.S. Air Force Combat Support in a High-End Fight*, RAND Corporation, RR-A316-1, 2021. As of February 15, 2023: https://www.rand.org/pubs/research_reports/RRA316-1.html

Thomas, Brent, Mahyar A. Amouzegar, Rachel Costello, Robert A. Guffey, Andrew Karode, Christopher Lynch, Kristin F. Lynch, Ken Munson, Chad J. R. Ohlandt, Daniel M. Romano, Ricardo Sanchez, Robert S. Tripp, and Joseph V. Vesely, *Project AIR FORCE Modeling Capabilities for Support of Combat Operations in Denied Environments*, RAND Corporation, RR-427-AF, 2015. As of February 15, 2023: https://www.rand.org/pubs/research_reports/RR427.html

Tripp, Robert S., Lionel A. Galway, Timothy Ramey, Mahyar Amouzegar, and Eric Peltz, *Supporting Expeditionary Aerospace Forces: A Concept for Evolving the Agile Combat Support/Mobility System of the Future*, RAND Corporation, MR-1179-AF, 2000. As of February 15, 2023: https://www.rand.org/pubs/monograph_reports/MR1179.html

U.S. Department of Defense, *Summary of the 2018 National Defense Strategy of the United States of America: Sharpening the American Military's Competitive Edge*, 2018.

Williams, David R., *Examining EXPRESS with Simulation*, thesis, Air Force Institute of Technology, Department of the Air Force Air University, 2012.

Yip, George S., "Global Strategy . . . in a World of Nations?" *Sloan Management Review*, Vol. 31, No. 1, Fall 1989.